# Juggernaut Drivers

*By the same author*

Truckers North Truckers South

# *Juggernaut Drivers*

LESLIE PURDON

Old Pond Publishing

First published 2006

ISBN 1-905523-59-9
978-1-905523-59-7


A catalogue record for this book is available from the
British Library


Published by
Old Pond Publishing
Dencora Business Centre
36 Whitehouse Road
Ipswich
IP1 5LT
United Kingdom

www.oldpond.com


Cover design by Liz Whatling
Front Cover painting by Alan Spillett
Printed and bound in Great Britain by Biddles Ltd, King's Lynn

# Contents

# Acknowledgements

I would like to thank my wife Pauline for all her hard work and encouragement with this manuscript.

I would also like to thank the road transport artist Alan Spillett for the painting used on the front cover. Alan and I go back nearly forty years. We were both employed at Arnold Transport, Gravesend, driving Atkinson lorries all over the United Kingdom. He is a genuine friend.

Last but not least I would like to thank Roger Smith of Old Pond Publishing for keeping me on the right road.

# Foreword

If you are looking for a man who is the salt of the earth then look no further than Les. Anyone who meets him will always be greeted with a smile, unless you happen to be someone who has knocked him in the past!

Les has been around transport all his life and can tell some hilarious stories as his previous book *Truckers North Truckers South* displays.

This time Les has turned to fiction, although I think the stories may be based on fact but with the names changed to protect the (not so) innocent! Be it fact or fiction, this is a very entertaining read with an insight into the transport world of the 1970s and 80s.

I wish Les all the best with this book and will always consider him a friend and colleague in transport.

TONY BEADLE
Road transport and shows manager
*Classic & Vintage Commercials* magazine
Kelsey Publishing, Kent

# *Dedication*

To Pauline
A transport driver's wife at her best.

# Chapter 1

# *Leyland Marathon*

I FOLLOWED Benny as he pulled into a lay-by. As I applied the brakes, they made a loud hissing sound. I climbed down from my cab and walked towards his truck. We went round to the offside of his rig and sat on the bank recovering from our hangovers.

It was a beautiful day. The air was fresh; the smell of lake Taupo mixed with the newly cut pine on the trucks was heavenly. While we sat there, the trucks made clicking sounds as the engines and exhausts cooled off.

As we looked out into the distance, I glanced at Benny. 'It's so picturesque and peaceful here. It's got to be better than Scotland.'

Benny sat upright, speaking in his high-pitched voice. 'You can't say that. Nothing could be better than Scotland. You're a saucy bastard.'

'Well, when I last went to Glasgow, I didn't think it was anything special.'

'Rich, all countries have their good and bad places.'

'Oh so now you're telling me the truth, Benny: Glasgow is a dump.'

'You're just trying to wind me up,' he replied loudly. We both lay back laughing.

We were tired and started to doze off. I opened my eyes and looked at Ben who was now asleep. I thought to myself, no one could have a better friend. He was a good man.

Ben suddenly opened his eyes. Perhaps we were sharing the same thought. 'Did we go too far?' I asked.

'Of course we didn't, Rich.'

He closed his eyes again. As I looked out across the lake my half-conscious mind was saying, 'Did we go too far? Did we?'

My mind started flashing back to when I was in the British army. I had reached the rank of corporal in the Royal Engineers. They knew us as Her Majesty's Gentlemen, but we knew ourselves as the ones with the bolts missing – in our heads, of course. Who would want to crawl across a minefield, pushing a bayonet in the ground looking for mines and feeling underneath them for booby traps?

Like many others in the sappers I became an explosives expert. As part of their professional and extensive training they taught me to wire detonators in pitch-darkness. It was there that I met this good man who became a firm friend: Benny Walker from Glasgow.

Benny was born in the Gorbals on the south bank of the river Clyde, which made his outlook on life different from mine. Benny was a man with a chip on his shoulder. He was very quick tempered and easy to wind up. Nevertheless, we did have something in common. Our fathers were truckers.

I was from the south of England, the Medway towns in Kent. As my name was Richardson they called me Rich.

On the day when I was demobbed after twelve years service I said goodbye to the lads in the NAAFI and then Benny walked with me to the main gate. We shook hands and embraced, both a bit emotional at parting. As I walked away, Benny shouted out, 'I never did like jumped-up corporals.' Without turning I waved my hand. I had a strong feeling that our paths would cross in the future.

Not long after leaving the army I got a job in Gravesend with Reed's paper mill, delivering reels for the daily newspaper. My truck was an Atkinson Articulator with a five-pot Gardner engine, a good old girl that never let me down. This job wasn't bad but I had a burning ambition to buy my own rig. The more I thought about it, the more it became an obsession.

Early one morning in 1970 as I drove up the Old Kent Road, I noticed many second-hand vehicles for sale. I walked up to the gate where a short stocky man with a ruddy complexion met me. Seeing my rig outside he thought I was going to ask directions. I told him that I was interested in buying a truck.

'Take your time and look around,' he said. 'I'll be in that hut over there if you need me.'

He had all makes: ERFs, Atkinsons and the most recent Seddon – but the one that took my eye was a Leyland Marathon that had the latest Leyland engine fitted with a supercharger. It was in remarkable condition with only fifty-three thousand miles on the clock. Although it wasn't the best looking truck there, I'd taken a shine to it. I knew it wouldn't need painting.

I went over to the old boy in the hut. 'How much for the Leyland Marathon?'

'It's a snatch-back,' he said, 'going cheap.'

This meant that the trucker who had previously owned it hadn't been able to keep up his payments so the finance company had taken it back. Most of the trucks in the yard were being sold cheaply for the same reason.

I looked at him through knitted brows. 'How cheap?'

He started rattling out the old patter. 'It's in very good condition all round. You can have it lock, stock and barrel, ropes and sheets included. The price to you, young man – no ifs, no buts – is three and a half thousand.'

'Start her up,' I said. 'Let me listen to the engine.'

No sooner had he turned the key, than she flew into life. It sounded good.

'I'll take it. I'll come back at the end of the week to pay for it.'

'You'll have to leave me a deposit,' he replied.

'But I've only got a fiver on me.'

He peered at me with beady black eyes. 'A fiver to you, young man, is a considerable amount of money. So give me that fiver. I know you'll come back because you won't want to lose your cash.'

He gave me a receipt for my five-pound note and I walked out of there on cloud nine.

As I continued up the Old Kent Road I knew that I was going to buy the rig, no matter what. I was getting excited by the mere thought that I was going to purchase my own truck but I was nervous at the same time.

I thought about the financial difficulties that the previous owner must have had. Because the truck had been snatched back it was still carrying his lorry sheets and the ropes were still inside the cab.

After I unloaded at the *Daily Mirror* I made my way back to Reed's. Going down the Old Kent Road again I was rubber-necking in an attempt to see the truck that I was going to buy.

When I'd parked up at Reed's, I wound the wheels down on the trailer ready for the night shunter to load. I thought to myself that it would be a good idea if I asked one of the fitters to come with me and have a look at the Leyland.

Jimmy Little, a good fitter, agreed to do it. 'You'll keep quiet about this for the time being, won't you,' I urged.

'Mum's the word,' he replied, pocketing the ten pounds.

Lifting up the boot of my car, I threw the trailer handle inside. You had to do this at Reed's, otherwise things had a habit of disappearing.

As I sat eating my dinner, I waited for my opportunity to tell Dad about my new venture. When I told him that I wanted to buy a truck, to my surprise he was interested.

'Well, Son, you're old enough to know your own mind. How much is this truck going to set you back?'

When I told him, I could see he was startled by the amount. 'My bloody house didn't cost that!' he said. 'How much money have you got towards it?'

'I've saved eight hundred and fifty from the army and Reed's. I'll have to get a bank loan for the rest.'

'Dennis, when your Nan passed away a couple of years ago she left us a tidy little sum – our nest egg for when I retire. We'll lend you the money on the condition that you pay us back. You've got five years with no interest. How does that sound?'

'I don't know what to say, Dad. I'm speechless.'

'Thanks wouldn't go amiss,' he laughed.

Mum put her arms around my shoulders. 'We're only too pleased to help you, son.'

It took me a week to sort out the insurance. When it was done, Jimmy Little and I jumped in my Austin 1100 and travelled to London. While I was going through the paperwork with the old boy in the hut, Jimmy checked the truck thoroughly. When Jim gave me the thumbs up, I handed over the cash. In no time at all I had the bill of sale in my hand.

As I was heading back to Reed's yard followed by Jimmy in my old car, the Leyland engine sounded grand. Barking along the A2, she was handling beautifully and I knew how good she looked on the outside. I was so proud.

## Chapter 2

# Chuckles

ON Monday morning I arrived early at Reed's and immediately reported to the transport manager's office. I apologised for leaving the truck in his yard and explained that I didn't have anywhere else to park it.

'Well, Rich,' he said. 'I've seen your truck and it looks good to me. I've got a delivery if you want it.'

'I'll take it, wherever it is.'

'It's for Thatcham. So while I'm making out the tickets, you go and get loaded.'

While the overhead crane driver was loading me, the lads shouted out, 'Who's bought his own truck, then.' I smiled.

When they had finished I threw the sheet over the load, but found that it only covered half of it. 'That's no problem,' they said in the office. 'Go and get one off the spare trailers and hand it back when you return.'

I turned left at the gate, the steering feeling good, especially after driving the Atkinson. I drove along the under-shore at Gravesend with the river Thames on my right. When I turned left up Pier Road, a very steep hill, the old engine started to bark. She was pulling great and I was chuffed to bits. I checked the mirrors to see if she was chucking out any smoke but the air

17

behind me was clear. I felt quite sure that in buying this rig I'd backed a winner.

I was now making my way towards Northfleet, the reflection of the Leyland looking smart as I passed the shop windows. My first day on the road was looking good, especially getting a payload from Reed's which was pure luck. The aim was to get to Thatcham – near Newbury in Berkshire – as soon as possible so that I could get unloaded and back in the same day. 'I've got to make this truck pay,' I told myself.

After leaving London I headed west along the M4. Checking the mirrors for wooden tops (police), I pushed the gas pedal hard to the floor and moved over to the middle lane. She was driving like a dream, the speedometer reaching sixty-seven miles an hour. Now that I knew what her capabilities were, I eased off the gas and continued the rest of the journey at sixty.

At Theale I turned off the motorway onto the old A4 for Thatcham, soon pulling in to Reed's main depot for producing corrugated cardboard. It was a good off-load. I drove back out of the gate and along the A4 for a short distance then pulled in at the Tower café. I sat near the window to eat my meal so that I could admire my rig in the August sunshine.

I was aware that I couldn't return to Reed's too early as it would mean cutting the job up – upsetting their own drivers who were on hourly rates of pay. That would put me in trouble with the shop steward who would ban me from entering the depot even though we were on good terms and I was a fully paid up member of the union. So I drove to Blackheath common where I contacted Reed's by phone.

The bad news was that they didn't have any more work for me, but told me to ring them regularly.

As I climbed up into the cab of the Marathon a thought suddenly occurred to me. I would drive to the Associated Portland Cement Works at Swanscombe, close to Gravesend, to see if they had any work. When I arrived there the transport clerk said that they had a load of special cement for oil rigs that had to be delivered as early as possible the next day.

'Go down to the yellow shed and report,' he said.

As I pulled alongside the shed, an AEC Mandator was already loading. 'Back up alongside him,' the charge-hand told me. The loaders started to load my vehicle from a conveyor belt.

As I stood there the driver of the other truck walked over. 'Are you going to Yarmouth too?' he asked.

'Yes, I am.'

As we stood talking, I could see from his manner that he was quite a character. I judged that he would be about ten years older than me, say forty-three. He was of small build, but had very piercing eyes; he also had a couple of teeth missing in the front. All in all he was a very likable but scruffy individual who looked as though he could do with a good wash. He was one of those people who laugh at their own jokes.

'My name is Charlie Sorter,' he said. 'But I'm known as Chuckles.'

'I can quite understand why.'

'Now and again when I wear a clean shirt and tie I like to be called Charles,' he laughed.

'Where are you parking up tonight, Chuckles?'

'Well, I wanted to go home, but I'd better not

19

because of the early delivery in the morning, and I don't want to overlay. So I was thinking of going up to Ipswich.'

'I can't go as far as that,' I said. 'I'll run out of driving time.'

At this Chuckles almost fell over with laughter, he was giggling so much. With both hands he clutched his knees, gasping for breath.

'What's so funny?' I asked.

'Driving time! I don't worry about things like that. I just keep going until I've had enough. It's bad enough trying to earn a living without worrying about driving hours.'

The loaders shouted out, 'Come on, lads, you're both loaded. Drive those trucks out of here.'

'All right. Keep your hair on,' I yelled back.

We roped and sheeted our loads. I used the small sheet that I had acquired with the truck, and it was just the right size for the cement. I was hoping that Reed's would forget the lorry sheet they loaned me as it would come in handy for high loads.

'If you're ready, Chuckles, I'll follow you,' I said.

We drove through the Dartford tunnel, along the A13 and headed for Brentwood. Soon we were punching down the A12 which was notorious for wooden tops. Chuckles certainly didn't hang around. He knew how to make the Mandator's 11.3 diesel engine sing – but it was belching out black smoke.

'He's taking a chance with that,' I thought. 'The rig doesn't look very healthy, either.'

Once we got the other side of Chelmsford we pulled into a transport café. Chuckles suggested we just have a cup of tea and a bun, as he knew of a nice little Chinese restaurant in Ipswich.

'Where are we stopping?' I asked.

'In the main lorry park opposite the football ground.'

'How can you sleep in an AEC, Chuckles, with a bloody great bonnet poking up in the air?'

'Don't ask, don't ask,' he giggled.

Just forty minutes after we rejoined the A12, we arrived in Ipswich and parked up. Chuckles told the old boy sitting in the hut that we would pay him when we came back from town. Walking away from the hut, Chuckles remarked, 'He's had that. I don't pay parking fees. We'll be long gone before he comes back in the morning. Having no names on our trucks has its advantages.'

I had to smile at him. He had the cheek of the devil.

We had a quick wash and brush-up in the gents. The meal in the Chinese restaurant was really enjoyable. As we ate I told him that being an owner-driver was all new to me.

'Don't worry. I'm an old hand,' he said. 'Been working for myself for five years – just about keeping the wolf from the door. Rates aren't that good just now,' he continued, 'but most companies pay on time, which is the main thing. We've not chosen the best years to work for ourselves. It's getting harder – owning your own truck isn't all honey.'

Chuckles asked me more about my plans. 'Where do you usually park your truck, Rich?'

'I haven't found anywhere at the moment.' I replied.

'I park up on my brother-in-law's smallholding in Rainham. He keeps a few chickens and so on. It's nice and quiet there so you can work on your vehicle without being bothered. The only trouble is that when you rev your engine the poor old chickens get so frightened that they fly up in the air. There's shit and feathers everywhere,' he said, grinning all over his face. 'If you fancy it, I'll have a word in his shell-like.'

'Yes, if he agrees I'll take you up on that, Chuckles. It's only a few miles from where I live.'

That night I had an awful sleep in the Marathon. 'I suppose Chuckles is used to it by now,' I thought.

Next morning I was woken by Chuckles banging on the door of my cab. 'Come on,' he was yelling. 'Let's burn rubber.' It was a quarter to five.

I climbed down from my cab, walked around the truck and had a good old stretch as my body felt as though it had been put through a mangle. I ached from head to toe.

When I jumped back behind the wheel and started her up I knew something was wrong. It took a long time for the air to build up, and the pressure started to drop. I immediately climbed back out and when I walked around the back I could hear air escaping. Chuckles heard the noise and came over to investigate.

'It's coming from your red airline connecter,' he said.

Chuckles climbed up at the back of the cab and disconnected the airline. Inside was a small rubber seal which looked like a tap washer. It was so badly damaged that it was beyond repair. 'We won't get one of these at this time of the morning,' he exclaimed. 'Your first option is to wait here until the garages are open at eight – but that means paying the lorry park attendant, and we don't want to do a silly thing like that.

22

'Or, I'll get under the back end of the trailer and let the air out of the tanks which will release the brakes. The only thing is, Rich, you won't have any trailer brakes. You'll have to drive to Great Yarmouth with brakes only on the unit, so you'll have to drive with extra caution, especially with twenty tonnes of cement up your rear.'

'I'll try that, Chuckles,' I said.

'OK. I'll drive steady.'

By the time I reached the other side of Ipswich, the brake drums had become really hot. Then I saw that I was going to have to go downhill towards Wood-bridge. I wasn't looking forward to this for one minute.

When we started down it was soon apparent that the brakes wouldn't hold me back and I was getting closer to Chuckles. The smell of the linings from the brake drums was drifting up into the cab. I started to panic, being so close to Chuckles's back end. I had to make a quick decision.

I decided to run the nearside of the truck against the kerb to slow her down. When Chuckles saw this in his mirror he immediately took evasive action, hitting his gas pedal hard. I was so relieved when the distance between us got wider.

When the road levelled off I continued driving along at a steady pace so that the drums could cool down, then pulled into a lay-by behind Chuckles. He climbed down from his rig and looked up at me quizzically. 'My word, you do look pale, Rich. I wonder why?'

'You cheeky bastard. I was shitting myself back there. I feel as though I've been on a suicide mission.'

'You've only got to stick it out until we reach Great

Yarmouth. You'll be all right. If I see any hills I'll flash my rear lights, put my foot down hard and let you get on with it.'

'You treat everything as a joke. Why?' I replied.

'Because, Rich, you're a long time dead and life gets tedious, don't it.'

I wasn't too happy about leaving the lay-by. Then I gave myself a good talking to. 'Dennis, my boy, do yourself a favour and relax. You're in East Anglia, not the mountains of Wales. If you can play silly buggers on a minefield, driving this rig has got to be a push-over.'

Being early morning there wasn't much traffic on the roads. This made life a lot easier. When I drove up hills, I really made her tramp but going down again was a nightmare. As I started to descend any appreciable slope I changed down into second gear, the engine scream-ing. I knew that I could easily knock out a valve which could go straight through a piston and damage the engine, but this was the only way I could control my speed. I eased my foot gingerly off the foot brake to allow the brake drums time to cool off.

As soon as the road levelled off I changed back up through the gears. I could hear my old Leyland saying, 'This is better.' The cool air was blowing through the radiator grilles.

What I did back there was bloody stupid, I know. But when you own your own rig you have to make a decision, rightly or wrongly. It's different when you're working for a company because all you have to do is pick up the phone, and help is on its way immediately.

By now I had caught Chuckles up. I followed him over a bridge, and drove alongside the misty river Yare for a couple of miles before we turned off right, eventually pulling into an American company. There were at least fourteen Yanks in the yard who all helped unload us. They seemed to have appeared from nowhere. We had a laugh with them as they shifted the forty tonnes which was on the two trucks. Chuckles and I helped by sliding the bags to the edge of the trailer. The perspiration rolled off us even though the temperature was quite cool in the early morning.

As we drove out of the gate heading towards the bridge, Chuckles stopped alongside the river. We stood chatting for a while, deciding what to do.

'I'm not going to look for loads around this part of the country,' Chuckles said. 'It's not worth it. I could have to go further from home and in any case the rates will be low because they'll be coming from a clearing house. I think it would pay me to get back to Kent. Anyway, I don't fancy two nights on the trot sleeping in an AEC. So it's up to you, Rich. What do you want to do?'

'I want to go home too.'

'OK. We'll make our way down to the Lowestoft Transport Company and get your airline sorted out. You should be all right now you've no weight on.'

It wasn't long before we reached our destination where the fitter was a young lad. As he pushed the rubber ring into the air socket line, Chuckles closed the air valves underneath the trailer and we heard the air rush through the lines. The air gauges in the cab went up very quickly.

'A two-minute job caused all that bloody aggro,' I said.

'Yes, that's road transport,' replied Chuckles.

On the way back down the A127 we pulled into the Half Way café for dinner. After a few phone calls Chuckles managed to get loads for us out of Honix Wharf in Rochester.

On our arrival to load we were told it was pulp, which meant that there was no roping or sheeting to do. 'Great,' we thought. 'Less work for us.' They gave us two loads each which had to be delivered to Aylesford, near Maidstone.

Afterwards I followed Chuckles to his brother-in-law's smallholding in Rainham, north Kent. We drove through Gillingham, and onto the lower road. Just as we approached Rainham we turned off left down a track on marshlands. We passed the old farm cottage where Chuckles and his family lived and twenty yards farther on pulled alongside a large black shed which was crammed full with auto jumble.

'Let's go and have a word with Len,' said Chuckles. 'We get on like a house on fire, which is just as well as we're married to sisters and he's also my partner.'

We walked farther down the lane past another old cottage 'That's where Len and his wife live.'

Len himself was in the field close to his cottage, surrounded by chickens, ducks and geese. He had cows grazing in the adjoining field and there was a distinct smell of pigs.

'Chuckles,' I whispered. 'I don't like the way the geese are hissing at me.'

'Oh, don't worry about them. They're just warning you to stay away. They'll get used to you being around after a while.'

'Hello. My name's Len,' said Chuckles's brother-in-law, proffering his hand. 'Sorry it's so dirty.'

Len was dressed like a typical farmer. You could see from his slightly bent back that running a farm had taken its toll over the years. His complexion was really weathered. His smallholding was some eight acres of poor soil near the marsh but he just about made it pay with vegetables and the livestock. I found him to be a very friendly and genuine sort of person.

'Can my mate park up here?' Chuckles asked. 'He's got nowhere to leave his truck.'

'No problem,' said Len. 'There's plenty of room. But I hope you'll keep the noise down. It frightens the chickens and puts them off laying.'

Chuckles laughed at this. 'As if we'd frighten your chickens!'

We walked up the lane which led to their farmhouse where I was introduced to Chuckles's wife, Jean. Her oval face had a beautiful complexion with dimples in her cheeks which showed when she spoke. She immediately put on the kettle for a cup of tea.

Jean and Chuckles's two children, a boy aged eight and a four-year-old girl, were playing in the back garden. On hearing their dad they came running in and Jean introduced me to them as Uncle Rich. The daughter, Jenny, came and stood by me, so I picked her up and put her on my knee where she clung to her dolly like grim death.

'What's your dolly's name?' I asked.

'It's Dolly, of course,' she replied, looking straight into my eyes.

We all laughed at the way she said it.

'That doll goes everywhere with her,' said Jean. 'She won't let anyone else hold it.'

She went on to talk about the farm. 'My sister and I were brought up on the land,' she said. 'This poor marsh soil isn't enough for us all to earn a living—'

'So that's why I bought a rig and we went into transport,' said Chuckles.

I thought Jean was a lovely person, the salt of the earth.

After a while I made my excuses and left.

# Chapter 3

# A Chance Meeting

I DROVE the Leyland for three years. But she was never comfortable to sleep in and over that period trucks with day cabs were disappearing fast as sleeper cabs came into force, doing away with the need for transport digs. At the same time drivers' transport cafés were closing while in their place Little Chefs were springing up everywhere.

Volvos hadn't been in the country very long but were now appearing in greater numbers heralding what was to prove to be a Scandinavian invasion. When I asked a few drivers what they thought of these trucks, their replies were always the same: 'Volvos are reliable and great to drive.'

So in 1973 I took a chance and invested in a brand-new Volvo 88 sleeper cab.

The part-exchange deal I received when I traded my Leyland for the Volvo was excellent. I was sad to see the Leyland go because she was so reliable and never let me down. She will always be my favourite truck – but all the same I felt I had to move on. I kept my old trailer because it was in very good condition.

My first load was out of Poplar, in London. I pulled off their weighbridge with nineteen and a half tonnes of lard on board which was to be delivered to Manchester. Over the past two or three years I had become accustomed to the engine roaring in my ears and holding onto the steering wheel with both hands. But in this rig the engine was almost silent and the steering was so light that at sixty miles an hour I could control the wheel with the tips of my fingers. It was superb. It made me wonder where the pulling power was coming from. Unfortunately this trucker's dream put the British trucks fifteen years behind.

I tramped on up the M1 feeling on top of the world. Both mirrors on the side of the unit were curved which gave an excellent view. I could see right down to the end of the trailer. As I pulled into the Blue Boar, I suddenly thought, 'This is the first time I've driven a truck and not felt jaded.' It was like being in a different era.

No sooner had I drunk a cup of tea and eaten a sandwich than I was back in the saddle punching north. I drove into Trafford Park and tipped. I managed to get another load out of a clearing house at a rate of eighty pounds for delivery to Maidstone. I was really chuffed because the back-load covered the running and diesel expenses.

By now I had driven over my hours so I turned off the M1 and drove along the A45 for about a mile then pulled into a lay-by where other drivers were also parked up. A public house was just up the road in which drivers could have a meal and wash and brush up before retiring.

Jumping up inside the cab I drew the curtains all round, ready now for a good night's sleep. The bunk was much more comfortable than the one in my old

Leyland, so on top of the best day's driving experience I'd ever had I also got the best night's kip.

At four the next morning I was back on the road. I kept the speedometer at sixty miles per hour all the way to Kent.

I still felt fresh as I made my way home. I parked my truck outside my Dad's house and was having breakfast by seven. Afterwards I drove to Sharp's sweet factory in Maidstone, arriving there just after eight.

While the forklift drivers unloaded the pallets of flavouring from my truck I gave Reed's a ring. They didn't have anything for me, but I did manage to get a load out of Rugby Portland Cement, Snodland, to go to Margate.

However, things then took a turn for the worse. Just as I was driving onto the A28 a car overtook me, blowing his horn and flashing his lights. When it was safe to do so, I pulled up on the edge of the road to get out and investigate.

I found that one pair of wheels on the nearside of the rear trailer axle had completely burnt off two tyres, and the wires from them had wrapped all around the wheels. I had never seen anything like it. All the money I'd earned that week would be spent on replacing the tyres. I felt like crying.

I had only got one spare wheel on the rear end of the unit, so I immediately took it off and jacked the trailer up. The wires had to be cut on the tyres, to enable me to get at the wheel nuts. It would mean driving to Margate with one wheel on the axle – but *c'est la vie*.

With a great deal of effort I managed to get both wheels off the trailer, the sweat pouring off me. When I

31

examined the spare I found it was the wrong size to fit over my drum. I clenched my fists and started pacing up and down. The air was electric. I had no money; I didn't even have a tyre call-out card. I just didn't know what to do. My main worry was that I had left a trail of burnt-out tyres on the A299. On top of that I still had my load to deliver.

I locked the door of the cab and started on foot to Birchington. As I began to walk I could see an AEC Mandator coming in the opposite direction, and then as it got nearer I could see it was Chuckles's. I ran to the side of the road to flag him down.

'What's wrong?' he shouted.

'The spare won't fit.'

'What do you expect? It's a bloody foreign one,' he giggled.

I was so relieved to see Chuckles that I laughed with him. 'I don't know why I'm laughing, Chuckles,' I said. 'I'm in the shit.'

'I'll soon have you back on the road. Don't just stand there. Get the jack out of my cab.'

I watched as he undid the wheel nuts on his trailer.

'What are you doing, Chuckles?'

'You've got to deliver this load, right? The best way out of this mess is for me to take a wheel off each of my axles and put them on yours. I'll just have to drive my truck home with single wheels on the trailer.'

My rig was soon fully operational. As I drove into Margate I thought to myself, 'There aren't many men out there who would have done that like Chuckles. But he's one of the old transport drivers.'

When I arrived back at the farm I found Chuckles

working in his shed and singing at the top of his voice. When he saw me, he raised his hand and pointed to the side of the wall where I saw my two wheels fitted with Michelin tyres.

'They look all right, Chuckles. Where did you manage to get those fitted?'

'Oh, I put them on. When I bought this rig the driver who owned it before me gave me a lot of spares. I kept them in the shed; I knew they'd come in handy one day.'

'How much do I owe you?'

'Don't you worry your head about that. Just replace them when you can. If you give me the cash I'll only spend it. What we'll do, Rich – only if you agree, of course – is make a spare wheel carrier and fit it on your trailer.'

He then took the measurements from his wheel frame to make one similar for mine. Chuckles pushed the frame under the trailer, and then lifted it up onto some bricks and welded it to the chassis. He drilled two large holes into the metal and welded two large nuts on the bottom of each hole.

'When we put the spare wheel on, the bolts will go through the holes and screw onto the nuts,' he told me.

'You're a clever old bastard, aren't you?'

'I know,' he said with a glint in his eye. 'I've got a spare wheel you can have with a Dunlop tyre, I know there're not much cop but it would get you home.'

'Thanks, Chuckles. You won't lose by this. I'll make sure of that.'

'Rich, let's lock up and go and see Jean. I don't know

about you, but I'm gasping for a cuppa.' Then Chuckles gave me a sly look. 'By the way, I forgot to ask you. How did your lorry run today?'

'It was OK. Why?'

'Well, when you last parked up here I put twenty gallons of paraffin in your tank.'

'You did what?'

'You heard,' Chuckles answered. 'We have it delivered now and again for heating and farm purposes. It has to be mixed with diesel, though. Otherwise it won't work'

'You're a lad, you are, Chuckles.'

On the way home I called into a hardware shop and bought some black undercoat paint. The next day I cadged a wire brush off my old man and went back to the farm. In the distance I could see Len boxing up his cabbages – every day of his life was spent in the fields. I laughed to see his geese following him around.

I unbolted the spare wheel from the trailer, wire-brushed the carrier and spare wheel, and then undercoated the lot. When I'd finished I slid the wheel back and tightened up the nuts.

If I hadn't painted it, it would have gone rusty. But in Chuckles's eyes that would not have been important. In no time at all I was banging on his door, feeling quite proud of what I had just done in so little time.

'What we'll do now, Rich, is go straight to Honix wharf to see if they have any work for us,' he said.

I followed Chuckles through Gillingham. Past Chatham dockyard he really pushed his truck to the limits, so it wasn't long before we pulled up outside the security gate at Honix. There were trucks everywhere:

Maylens, Arnolds and Swain's Transport, they were all waiting to load. As I sat in my cab behind Chuckles I thought to myself, 'Please let there be some work for us.'

We eventually reported to Mr Burley in the transport section who told us that two new orders had just come in. One load was for Reading, the other for Bristol.

I looked at Chuckles and said, 'I don't care what I do or where I go.'

'Well, Rich, I'd sooner go to Reading because by the time we finish loading it will be near dinner time. If I go to Bristol I'll have to spend the night in the AEC, and I don't fancy that.'

'OK by me, mate,' I said. 'Bristol here I come.'

Looking at my delivery notes, I saw that my load would be reels of paper. I couldn't believe my luck. I handed my notes to the forklift driver at the paper shed, after which he loaded me straight away. I had to turn my large lorry sheet inside out because it had Reed written on it.

Once loaded I drove to the security office and had my delivery notes stamped. In no time at all I was driving back out of the gate, over Rochester Bridge and sitting in the Merry Chest on the A2 having an early dinner.

Driving through London wasn't too bad apart from a little traffic in Camberwell. I carried on to Vauxhall, through Nine Elms, past Battersea Dogs' Home then over Battersea Bridge and along the Embankment. After Earl's Court I eventually reached the M4. As I drove along, the sky was becoming very dark. Then all of a sudden the heavens opened and it rained so hard I had to ease off the gas. By now my speed was down to fifty and the visibility through my rear-view mirrors was really awful. The windscreen wipers had difficulty in coping with the violence of the rain.

By the time I had passed the Swindon turn-off the rain had stopped, the sun was shining and the roads were dry. 'This only happens in England,' I said out loud as I parked in a lay-by and checked my delivery notes. Although they had told me I was going to Bristol, the notes read Avonmouth. I had to find a place called Portbury and lost a lot of time doing so. By the time I had tipped and folded up my sheets it was half past five.

On my way back to Avonmouth I noticed a transport café, so I pulled in and had a cup of tea. As I sat daydreaming, a hand suddenly grabbed my shoulder and a voice said, 'I never did like jumped-up corporals.'

I knew right away even before turning around who the voice belonged to. I shouted out, 'Benny!'

I was so thrilled at meeting up with him again. We must have sat and talked for at least an hour. 'Where are you staying tonight?' he asked.

'I was going home to Chatham. But now I have a strange feeling that I'll be staying in Bristol.'

'Too right you are,' Benny exclaimed. 'I've a back-load for Glasgow tomorrow, so I'm staying here tonight.'

Benny followed me to the lorry park which was in the centre of Bristol. We parked up and then went off for a few jars and a bite to eat. We were soon reminiscing. 'Ben,' I said. 'Do you remember when we were on booby traps and I would pull a pin from a grenade and throw it to you?'

'Aye. Then I'd throw it as far away as possible. We were bloody mad in those days, Rich. I remember it as though it were yesterday.'

'What did you do when you got demobbed, Ben?'

'Oh, the usual things. Fell in love, tied the knot, but it didn't work out for us. We were too young, I suppose. So I'm single again, fancy free and living life to the full.'

'Do you own this truck?'

'No, but the owner is not in very good health so he's asked me to drive it for a month or so. Work-wise he's scratching the bottom of the barrel. I'll probably hang my boots up when I get back home. Though I've heard, Rich, that there's a lot of work in the south-east of England and the paths are paved with gold. . . .'

'Benny, if you believe that, you'll believe anything. Still, if things are a bit dodgy in Scotland, why don't you come down here? You'll feel quite at home because there are more Scots here than there are in Scotland. There are plenty of transport companies in Kent that pay good wages, and there's no shortage of digs. You could always give it a try.'

'Thanks. I might just do that,' he replied.

We both had an early start in the morning so we weren't late to bed. As I drew the curtains, undressed and got into my bunk I thought about this chance meeting. Then I drifted off to sleep.

The following morning I quickly made up my bed in the cab, climbed down from my rig and banged on Benny's cab door to wake him and say goodbye.

'See you sometime, Rich. You take care now.' He waved to me as I drove off the lorry park.

I followed the M4 signs out of Bristol, the Volvo singing as I drove. Eventually I pulled in at Honix at seven forty-five that morning, which wasn't too bad.

Over a period of time, Honix had given me a great deal of work. I had been very lucky as money was coming in regularly. Also, Chuckles had been really good to me. Not having to pay rent to park my truck was a godsend. If either of the trucks had any defects we'd work all day Sunday repairing them.

I had become very fond of Chuckles, an all-rounder who didn't have a bad word to say about anyone. As he had been so good to me I decided to give him a Christmas box. What he had done for me in the past would have come to a lot more than I gave him. The way I looked at it, too, was I'd rather Chuckles have it than the taxman.

In my healthier financial position I thought about investing in a new car, as my old banger was well past its sell-by date.

Then I thought to myself, 'Rich, get your priorities right. I'm only rich by name.'

Now and again Mum would hint that it was about time I bought my own house. But I knew where my bread was buttered. I wasn't at home much anyway; most of my time was spent round at the farm with Chuckles.

One night my family and I were sitting in the front room watching television when there was a knock on the front door. When I answered it, to my amazement there stood Benny complete with suitcase. 'Well, this is a surprise,' I said, standing there completely dumbfounded.

'Aren't you going to ask me in?' Benny smiled.

We embraced. 'Of course. Come in, come in.' I introduced him to my parents.

As we sat eating our evening meal, Dad asked Benny if he would care to stay for a couple of days until he found somewhere permanent. 'I'd be delighted,' replied Benny. 'If it's no trouble.'

Dad, being Dad, said, 'I wouldn't have asked if I minded, young man.'

After dinner I drove to the farm with Benny as I wanted him to meet Chuckles and his family. As usual Chuckles just walked into the kitchen with muddy boots and up to his eyes in grease.

Jenny came in. Since we arrived she hadn't taken her eyes off Benny. I think she found his accent fascinating too. I had told her his name was Uncle Haggis. As the evening wore on she became bolder, calling him Uncle Haggis and speaking to him.

After an enjoyable evening we made our way back to my parents. Driving back I glanced over at Benny and said, 'Who's Uncle Haggis, then?'

'There you go again, winding me up,' he said, with a sparkle in his eye.

# Chapter 4

# *The Beginning of 'North Kent'*

BEN loved the Volvo. For the first three days he rode shotgun with me but after that he did most of the driving.

I told him that if he worked for a company for a short while it would tide him over and enable him to get a deposit for his own truck. He really didn't like the idea of working for someone else, even though he knew it made financial sense.

'You realise,' I said, 'that if you do buy your own rig, it's going to be bloody hard work – and at times there won't be much work around.'

'Oh I know that,' he replied. 'But don't you go worrying about me. I'm a Scot, I'll get by.'

The following Saturday morning Benny helped service the trucks. When we finished I took him for a drive up to London and before long we were going down the Old Kent Road. I showed him where I'd bought my first rig. Benny got all excited. 'Let's have a word with him now,' he declared.

The old boy recognised me. 'We've come to look at some of your old snatch-backs,' I said.

'Well, I've got an A Series ERF that I think may interest you.'

While I was talking to the old boy, Benny had a look over the truck. 'It's not a bad rig,' he said. 'How much do you want for it?'

'How does three and a half thousand sound?'

Benny started to haggle about the price. It was when Benny called him a crook that I walked away with embarrassment. I thought back to when I'd bought my own rig without any problem at all. But, as I've mentioned before, Benny had a very quick temper. When he eventually knocked the seller down to two thousand pounds they shook hands on the deal.

I walked over to Benny and said, 'It has got an engine, I hope. If not, it won't start. Is the back axle noisy? Does it jump out of gear?'

'How do I know? I haven't driven the bloody thing yet,' he answered, getting niggled again.

'Well, you're the one buying the rig, aren't you? Oh, and another thing. You do realise it hasn't got a sleeper cab, don't you?'

By this time Benny was getting really irate. I loved winding him up. He went into everything like a bull in a china shop.

Benny walked over to the owner and asked him if he could take the ERF out for a test run. Ben started her up. The Cummins engine gave a crackle as he drove out of the gate. I liked his care-free disposition – but I had to laugh. He had no tax, no insurance and no trade plates. It wasn't long before he was driving back through the gate. The truck looked respectable and it sounded good too.

As he climbed down from the cab, Benny said, 'I like it. It sounds as sweet as a nut.'

'The only trouble is, Ben—'

'Now what?'

'Just a small thing. Have you thought about a trailer?'

The old boy overheard us. 'I'll do you a deal,' he said, 'if you're interested.'

'I like a laugh,' Benny replied. 'Let's have a look at this trailer, then.'

The owner pointed to the corner of his yard. When we looked we saw the funny side of it: the trailer was an aircraft carrier known as a 'Queen Mary'. It was about sixty feet long and over thirty years old. 'Bloody hell, mate. You're having a laugh aren't you!' Benny exclaimed. 'If you put more than ten tonnes on that, it would break in two.'

'Come with me, lads, I've got another over there. It's in very good nick.'

When we looked it was a Scammell. By this time Benny was getting very agitated and I could see he was losing his rag again. I was right.

'Listen here, you ponce,' he said. 'It's a singe axle and can only carry fourteen tonnes. If you're trying to treat me like an idiot then you can stick this unit up your arse, and we'll go somewhere else.'

'Don't be like that,' the old boy said. 'Come round the back and I'll show you something different.'

When we saw this trailer we both knew it would be the one: a Crane Fruehauf, plated and with Michelin tyres in excellent condition.

'How come it's got a German name?' Benny remarked.

'Here we go again,' I thought.

'I know it's a German-sounding name, but Crane is an English company,' the old boy replied. 'The trailer was made in England.'

By this time I was getting a bit impatient, so I stepped in and asked, 'How much?'

'Two thousand pounds.'

I could see Benny's ears twitching. 'I'll tell you what, laddie,' he said. 'I'll make you a deal. The unit and trailer for two thousand one hundred and fifty pounds. How's that?'

I walked away. Benny was beyond a joke. I stood leaning on the gate post and listened to their haggling. I didn't really want to get involved. They argued for what must have been half an hour. 'Any minute now he is going to tell Benny to get off his property,' I thought, but luckily the old boy didn't. I must say, I couldn't have coped with a customer like Ben.

Then things started to move. Benny walked round, jumped up into the cab of the ERF, drove back on to the trailer and hooked up. He wound the wheels up on the trailer, connected the air hoses, revved the engine to fetch the air up, then drove out of the yard and up the Old Kent Road. When he returned, Benny backed the lorry and trailer into the corner and told the old fellow he would be back in three days.

By this time I was so exasperated that I walked over to the owner and asked him how much deposit he wanted.

'Well, if I remember rightly, the last time you bought a rig off me you only had a fiver.'

'Blimey, you've got a good memory.'

'You have to have in this game. But this time,

because of all the aggro I've had from your Scottish friend, I want at least a thousand pounds.'

'How much is it overall?'

'Two thousand nine hundred.'

'Ben, you've done well there, my boy,' I thought. He had managed to knock him down by one thousand one hundred pounds. I wrote out a cheque for a thousand pounds and told him that we would be back in three days time to collect the rig.

As I drove through Deptford, Benny said that he hadn't been expecting me to pay the deposit.

'It's a pretty poor show if I can't help a friend,' I replied. 'Look, why don't you let me pay for the truck outright so that you save paying interest on a loan. That's phenomenal at the moment.'

Benny did protest but after a while he could see it made sense. The relief on his face was a picture.

We chatted all the way to Gillingham, mainly about the old days. As I drove up the lane towards Len's farm, I glanced over at Benny. 'I haven't got any doubts about Len letting you park your truck there,' I said. 'He's a hard-working, decent sort of chap. A family man.'

I parked the car, then Ben and I walked over to Len who was cleaning out the chicken coops. The stench was a bit strong. As soon as Len saw us, his eyes lit up. He immediately threw the fork into a pile of dung and with large strides walked towards us, smiling. 'Len, I'd like to introduce you to a great friend of mine,' I said. 'He's just bought a rig.'

They dispensed with the formalities as Len's hands were really grubby and he didn't smell too wholesome either. 'Len, I wonder if I could ask a favour of you. If it's no, I'll understand.'

He looked at me quizzically. 'You're going to ask me if Ben can park his truck here.'

'Well, yes.'

'Any friend of yours is a friend of mine, so that's no problem at all.'

Ben thanked him profusely, and said we would see him later.

As we headed back towards my truck we decided to give the old girl a wash down. The pair of us got cracking and in a couple of hours it was done. It looked great when we finished.

It was now late afternoon and darkness was falling, so we made our way home for a bite to eat. We were ravenous.

After our meal we sorted out the insurance, tax etc., for the truck Ben was going to buy. He was getting really keyed up about owning his own truck. That's all his conversation consisted of, he was so keen. I did impress on him that he wouldn't have much money for a long time.

'I've thought about that a lot,' he replied. 'But by hook or by crook I'll make that old truck pay. After all, it's going to cost an arm and a leg.

'Rich, I've been thinking,' he continued. 'When I get my rig, I'll spray it.'

'It's not that easy. You can't just pick up a spray gun and do the job. It's a skilled worker's job.'

Ben laughed. 'I've done quite a bit of spraying, laddie. I'll take the doors off Chuckles's shed, drive the unit in, mask the windows, mirrors and engine, and then spray the cab and call it a day. The next day I'll back her out, reverse into the garage and spray the chassis. After that I'll hand-paint the wheels.'

'What will the finished job look like?' I asked, trying to keep a straight face.

'Like a brand-new truck, of course, you saucy bastard.'

An idea suddenly struck me. 'Ben, if you're as good as you say, paint your rig and Chuckles's truck black and white the same as mine. Then the only thing we'll have to do is give ourselves a name. How about calling ourselves the North Kent Carrying Service?'

'Nah, that sounds as though we're delivering bloody parcels.'

'I suppose you're right. What do you think about North Kent?'

'You've got it, Rich. It's short and sharp.'

Ben went on to say that when he was living in Glasgow some drivers put a lot of their hard-earned cash into their businesses. Then overnight they went bankrupt and the names of their companies went with them. So his philosophy was never call a business by your own name.

The following day Benny and I drove to the farm to do some maintenance on our trucks. It pleased me that Benny had volunteered to help.

While we were working I asked Chuckles what he thought about trading under the name of North Kent and painting all the trucks the same colour. Chuckles thought both ideas were excellent. 'My old mate Peter Darland will lend us his steam cleaner,' he said. 'That will save a lot of time.'

'In that case why don't we do all three trucks and trailers properly,' said Benny.

Chuckles eyed us both and said, 'I've got ten gallons of red coach paint. I fell over it, didn't I, a few years ago.'

'That's settled then,' I piped up. 'Red it is.'

'In the meantime,' said Chuckles, 'I'll get organised and clear the garage out. I'll remove the doors and hang an old truck sheet in their place so when you're spraying, the wind won't blow in. We'll do the first job next Friday. I'll jack the truck up and take all the wheels off the unit. Rich and I will paint the wheels by hand, then Ben can spray the cab. On Sunday Ben can spray the chassis, axles and springs.'

'That's fine by me,' said Ben.

'Steam cleaning is a horrible, dirty job,' Chuckles went on. 'We haven't got proper ramps so as soon as you and Benny collect his vehicle from London, I'll steam clean it and check the wagon and trailer for defects. Then I can remove the wheels and Benny can drive my AEC. When the truck is ready for painting we will swap over again until we have done all the trucks. It's got to make sense, lads, because after steaming I've only got to walk a few yards before I'm indoors. I can be in a hot tub before you can say Jack Robinson.'

Benny and I agreed that this was the best idea. 'North Kent' was becoming a reality.

The following Monday morning, we banged on Chuckles's door. The bedroom window opened and he appeared, all bleary-eyed.

'It's half seven,' I shouted. 'About time you were up.'

He threw the front door keys down and we let ourselves in.

Chuckles finally put in an appearance, smiling all over his face as usual. 'How about that for good timing,' he said, eyeing the teapot.

I poured tea into a pint mug and handed it to him.

Chuckles started ringing around for work about eight. We were all concerned as there didn't seem to be much around. After a few phone calls Chuckles said, 'I know what, I'll try Atlas Stone in Strood. They might have some work.'

After what seemed like an age, Chuckles finally got through. He told the fellow there that we had two thirty-two tonners, with a carrying capacity of forty tonnes. We saw from Chuckles's face that the answer was yes.

'We've got to get there straight away,' he said. 'The load is for Worthing in Sussex.'

I've never known anyone to get washed and dressed so quickly. When we reached our trucks, we climbed aboard and started them up with a roar as the air built up into the tanks to release the brakes.

As we moved off, Chuckles gave us the thumbs-up sign. He led us out of the gate and along the lower road into Brompton, past Chatham dockyard and through Chatham High Street, which led into Rochester High Street. Then we drove over Rochester bridge which brought us into Strood, where we drove down Commissioners Road and into Atlas Stone.

The stone was loaded by a forklift, twenty slabs at a time, squeezed together by a clamp. This meant there was no work for us and within an hour our trucks were loaded and ready to roll.

As I drove out of the yard I realised that for the first time my Volvo was making groaning noises. The dead weight was making the spring hangers and floorboards creak. When we stopped at the lights in Strood High Street, Benny and I could hear Chuckles's trailer grunting and groaning in front of us. It sounded like an old galleon on the high seas.

Going through Southborough, the last shop on the right was a transport café where we stopped and had breakfast. While we sat, Chuckles said how the weight had affected his pulling power when he went up past West Malling aerodrome. Although my old girl wasn't overloaded and the weight – twenty tonnes – was right, I felt the same.

Chuckles explained how different loads of the same weight affected trucks in different ways. With a normal load the momentum of the truck pushes you forward a little when you change down going up hill. However, with slabs, iron, lead and bricks, the load doesn't push you forward; it wants to go backwards with gravity. This is what is known in the trade as a dead weight.

After breakfast it wasn't long before we were driving through Tunbridge Wells on the A26 heading towards Brighton. Although Brighton was only fifty miles away it was a long, hard slog, especially driving up Crowborough Hill.

The worst was to come because we had to drive up the really steep hill in Lewes town centre. 'Any minute now the slabs will lean backwards, and if that happens the rope at the back will break,' I thought to myself. Lady luck was with me that day and my load stayed upright.

By now we were driving on the level. Chuckles's truck was billowing out thick, black smoke as we both tramped on towards Brighton. I followed Chuckles along Brighton front, and within forty-five minutes we were in Worthing. As we approached the railway crossing Chuckles turned left and about three hundred yards down the road we pulled up alongside the kerb.

There were two men to each trailer and when they had unloaded about ten slabs we were told to move our trucks up. They unloaded Chuckles and me at the same time.

As we drove back out of Worthing, I shouted to Benny, 'Wouldn't it be great if all loads were as easy as this?'

'It would be lovely. But we'll have to take the rough with the smooth,' he answered.

On the outskirts of Brighton, Chuckles pulled up at Ernie's café, in the middle of a parade of shops. After having a mug of tea, the three of us walked to the other side of the road where our trucks were parked.

'You can drive, Ben,' I said.

'No,' Chuckles said. 'He can drive mine. Let him see what it's like to drive a man's truck.'

I mentioned to Chuckles that his truck was billowing out a lot of smoke. He told me he had noticed, but it would have to wait as money was a bit tight. From the time we left Brighton to where we stopped at Snodland the journey took about two hours.

Chuckles rang Atlas Stone again. They told him that

we could come back and load our trucks ready for tomorrow. The load they gave to Chuckles was for Worthing; mine was for Hammersmith in London, which made a change.

We drove our trucks back to the yard fully loaded with paving slabs. By now I had become firm friends with Chuckles's brother-in-law, Len. He was the kind of man who would never ask me for money for parking my truck, but although I wasn't a family man myself I knew he must find it hard-going bringing up children on a low income. I used to give him some cash and I tried to look out for other ways to help.

On one trip I was delivering to a chemical factory when I noticed at least forty plastic barrels thrown to one side. I asked one of the managers if they were being chucked away.

'If you want them they're yours,' he said. 'It'll save me the trouble of getting rid of them.'

I drove back to the farm, parked up, quickly jumped from the cab then shouted across to Len who was feeding his animals. 'Are these any good to you?' I asked, pointing to the barrels on my trailer.

When he saw them his face lit up. 'You bet!' he answered.

Chuckles burnt the tops off the barrels and Len wrote 'pig swill' across them.'

Next day he started delivering them to canteens at local schools and factories, getting himself some free food for his livestock.

Benny's documents arrived in the post the day before I was due to go to Hammersmith. Early the following morning we were punching down the A2 towards

London, no more than a whisper coming from my Volvo.

The truck had an eight-speed gearbox that was a 'range change'. As I went up through the gears to fourth, I pushed a small lever on top of the gear stick then went through the same gears again – so the truck had four low gears and four high. It had taken me some getting used to but once I mastered it, it was fine. Going down Blackheath hill I moved my foot to the middle of the floor and pressed the exhaust brake. The way she kept the same speed all the way down was brilliant and I would have loved to have had this exhaust brake when I had trouble with the Marathon driving to Great Yarmouth. This Volvo was certainly a dream.

Very soon we were making our way west for Hammersmith. We arrived at seven fifteen and the Paddies turned up at seven forty-five. Once again, all we had to do was move the truck up so that they could unload the slabs along the road. As soon as we were empty Benny shouted, 'Old Kent Road, here we come!'

As I drove through London I named the places that we were driving through for Ben so that he could learn the shortest way from the west of London to Black-heath which then leads to the A2 for Kent.

'I'll tell you what. How about going past a clock that's named after you,' I said to Ben.

'You don't mean "Big Ben" by any chance?' he asked with a big smile on his face.

I drove along the Embankment, around Parliament Square to Big Ben, then went over Westminster Bridge through the Elephant and Castle and on to the Old Kent Road.

As I pulled up alongside the kerb I said, 'I can't wait here long. I'll get a ticket. If I'm not here, turn right out of the yard and keep on that road until you reach Blackheath common. I'll be at the tea stall.'

At the common I treated myself to a mug of tea and a roll. After about an hour I was becoming a little agitated as there was no sign of Benny. I began to wonder if everything was all right. Had he broken down, or run out of diesel on the way here? What shall I do? I asked myself.

As I sat there I heard the roar of a Cummins engine: Benny in the ERF. I felt the relief wash over me. I got up and walked across the green to his truck which was pulled up behind mine.

'What happened? I was getting concerned about you,' I said.

'Well, I had to check the truck and make sure he hadn't pulled a fast one by swapping the tyres or trailer over.'

I shook my head in disbelief.

'Rich, I don't trust no bastard. I learnt that much a long time ago.'

While Ben fetched himself a mug of tea and a cheese roll I had a good look at the truck. It was clean and in very good condition.

'What do you think of it, Rich?' Ben asked.

'The old boy at the truck sales certainly has some good snatch-backs,' I replied. 'You won't go far wrong with this motor, The Cummins engine is very reliable. Just remember to keep your foot off the throttle as much as you can, otherwise you'll have a BP tanker up your arse all day to top you up.'

'They're not that bad, are they Rich?'

'No, I'm only kidding. I see it's got a David Brown straight-six gearbox and the engine is a two-twenty, with Kirkstall axles which are the best. You shouldn't have any trouble with this truck. This is the last model ERF made in the "A" series. Did you know that, Ben?'

'No I didn't,' Ben replied. 'I just thought it was a good truck.'

'This model was replaced with the "B" series to keep up with Scania and Volvo.'

'Fascinating,' said Benny, a bit sarcastically. 'So, do you actually like it, Rich?'

'Yes I do. You've got a good truck there and the tank's half-full with diesel. It's roughly eighteen miles to Gillingham from here, so you should just about make it without filling up.'

'You always have to spoil things by taking the piss out of me,' Benny said before we climbed into our trucks and headed back to the farm.

As soon as Chuckles saw the ERF he put a large bar between the spring and chassis and started to rock it up and down to check the play in the shackle pins and king pins. He agreed that it was a good truck. 'Just be careful when you drive up Swanscombe cutting fully loaded,' he said. 'You'll burn twenty gallons of fuel just to get to the top of the hill.'

Benny wasn't amused. He pointed to me: 'I know that bastard has put you up to this.' Then he stormed off, sulking.

Chuckles told me that Atlas Stone had already paid us

for some of the work we'd been doing. 'They're not the best rates, but it's better than a kick in the bollocks,' he said. 'We've got some more work, too,' he continued. 'Two loads for Sheerness council yard. If it's OK with Ben I'll steam-clean your truck and you can drive my AEC.'

Ben and I drove out of the yard and headed towards Atlas Stone. Ben was to be loaded with two-foot paving slabs and I had to be loaded with three-foot slabs. In just over an hour's driving time we had pulled into the council yard at Sheerness.

When I pulled alongside the loading bay a gang of workers walked the paving slabs off the side of the vehicle. After I drove out of the way they unloaded Ben's.

Back at the farm we couldn't believe how much Chuckles had done. The trailer was unhooked, the mudguards were off, and the unit only had two single wheels on it. Chuckles had taken the other two off each side and removed the wings. We could see that he got more pleasure working in the garage than being on the road.

The following day Ben and I managed to get a load out of Rugby Portland Cement in Snodland for delivery in Portsmouth.

By the time we arrived back at the farm it was late afternoon. Chuckles had just finished steam-cleaning Ben's truck. Well, we'd never seen anything so comical; he was smothered from head to toe in gunge, oil and water. He was blacker than black. All you could see of him were the whites of his eyes and the pink part inside his mouth. We stood there laughing so much that our

55

sides hurt. He looked down at himself and began to laugh with us, but when he laughed he couldn't stop.

After a while I controlled myself enough to say, 'We'll have to steam-clean you now, the state you're in.'

He knocked on the door and Jenny answered. 'Yes?' she said.

'What do you mean, "yes". I'm your Dad.'

'No you're not. My daddy's white.' She turned and shouted for her mum.

When Jean appeared she took one look at him and said, 'You're not coming in here. Go around the back and get undressed in the garden, I'll give you an old sheet to cover yourself up, then I'll run a bath for you.'

Chuckles grimaced as he walked off.

When he came back in his sheet we sang one of Al Jolson's songs: 'Mammy don't you know me, I'm your little baby. . . .'

The two of us sat at the kitchen table, talking to Jenny. Half an hour later Jean appeared, looking rather annoyed. She told us that she'd had to refill the bath with soda because Chuckles still wasn't clean.

When Chuckles finally came into the kitchen we just stared at him because this time he was as red as a beet-root. He was glowing like a lobster. He looked at Jean and said, 'You silly woman, you put too much soda in the bath.'

When I arrived back at the farm the following afternoon, Benny had finished undercoating the unit and trailer and had started to hand-paint the wheels. He said that he had learned to spray in the dry docks along the river Clyde. 'An old boy there took a shine to me and taught me how to use a spray gun properly. He had been spraying since he was a lad. For the top coat I'll heat the paint up and then spray the cab while it's hot. When it's dry it will have a high gloss and be as smooth as a baby's bum.'

'I'll give you a hand to paint some of these wheels,' I said.

'OK. But make sure you do it properly, Rich.'

Looking at Ben with a twinkle in my eye I said, 'Ben! I'm the corporal, you don't tell me what to do.'

The following day I loaded twenty tonnes of paper out of Gravesend which had to be delivered to Christchurch in Dorset. The rates were excellent. I made my way to the A2 and the Volvo cruised up Swanscombe cutting like a dream. By the time I reached the top my speed had only fallen to thirty miles per hour which wasn't at all bad. Once again I thought to myself what a superb truck this was turning out to be. It was one of the finest vehicles in the world.

I parked up outside the Merry Chest café. While I was drinking my tea I got talking to some of the local truckers. By now there were quite a few owner-drivers; a couple of them asked about my rig and what it was like so I gave them a few ideas of what to look out for when the time came to replace their own vehicles.

When I left the café I drove around the outskirts of Greater London and was soon passing through Croydon, then on to Mitcham, past the common. On the Kingston bypass I cruised along at sixty miles an hour heading towards Guildford, then on to the A31 and across the Hog's Back. Now I was well on my way to Southampton, I turned right onto the A35.

When I pulled into Shand Kydd's wallpaper factory,

I didn't feel jaded at all. If I had been driving my old Atkinson I would have fallen out of the cab, because it took all my stamina to drive. The noise from the Gardner engine had been unbearable and that was why the old transport drivers suffered from tinnitus in the ears, a complaint that would not go away.

Within two hours I was pulling out of Shand Kydd's and making my way to the Towers café in Basingstoke where I parked up for the night. I felt very satisfied with myself because I had loaded and tipped in one day. I knew that with this rig I could get an early start in the morning and miss all the traffic. It would take me two and a half hours to get back to the farm whereas it would take me four hours if I travelled on now and caught all the commuter traffic that night.

I had a lovely meal in the Towers and a comfy bed in the cab. What more could a man want?

Just before retiring for the night I gave Chuckles a ring to see if there was any work. He had rung around and managed to get a couple of loads out of the cement works which had to be delivered to Great Yarmouth. We arranged to meet at Associated Portland Cement in Swanscombe at six the following morning.

'By the way, Rich,' Chuckles went on to say. 'You know you told me about your parents getting edgy when Ben moved in with them, which of course I can fully understand.'

'Yes, that's right.'

'Well, Len says that Ben could stay with him and his wife, as they've got plenty of room. At least it will tide him over for a while until he finds somewhere else to live.'

I knew my mum would be pleased about that. She'd also started nagging me and saying that it was about time I got married and settled down. In other words she was really telling me that it was about time I flew the nest.

## Chapter 5

# Recession Creeping In

WHEN I woke at three forty-five it was drizzling with rain. I quickly got dressed and was soon driving along the A3. It was very dark and I kept my speed at a constant sixty-five miles an hour.

I arrived at the Swanscombe cement works at just the right time, six o'clock. It was now daylight and had stopped raining. I drove onto the weighbridge where I was handed our delivery notes. Chuckles by now had pulled up behind me.

We backed onto the loading bays. While I was being loaded I took the opportunity to make my bed in the cab, and then use their facilities to wash and brush up.

As we walked back, Chuckles started chatting about how good Ben was with the spray gun. 'Rich, I can't wait for him to start on my truck.'

'But this time, Chuckles,' I replied, 'we'll organise things properly. We'll wear the correct clothing: water-proof suits, goggles and Wellingtons. We'll have to make a proper ramp, too.'

'You're right, Rich. It was silly of me. I must have been away with the fairies that day.'

'Away with the fairies? You're lucky it wasn't a wooden suit,' I laughed.

Chuckles then started talking about the journey to Yarmouth. 'I don't care what happens today,' he said, 'so long as I get back home. I can't sleep in the AEC. It's too uncomfortable.'

'We can always go into some digs, if you like,' I said.

'No. I'm going home.'

When we left the yard with Chuckles in the lead he drove his old truck as if there were no tomorrow. There was no doubt about it, his AEC could hammer on.

I hung back on the hills because my truck was far superior to his. I honestly don't know how he'd got away with it for so long. It billowed out so much black smoke at times, it was pure luck that he hadn't been stopped by the wooden tops.

We had a good run through to Essex, stopping for breakfast en route. As we were eating we overheard a couple of drivers talking about some congestion on the other side of Witham which we later found out to have been caused by an accident. Apparently a truck had gone off the road and hit a bank on the near-side. The front wheel was jammed up against the fuel tank. The driver was unhurt and lucky to be alive. No one else was involved but the truck was a complete write-off.

'Come on,' I said. 'We'd better make a move. We've still got a long drive.'

When we came to the scene of the accident we saw that the truck was a fully loaded AEC from Barnsley. The off-side of the cab was a mangled lump of metal. There was a gaping hole where the vehicle had been pulled out from the bank – it must have gone in about three feet.

It was a good run through and we didn't stop again until we reached Great Yarmouth. Although there was no one at the dry dock, a telephone number was pinned to the office door. We walked into the unlocked room and rang the number which was answered by a man with an American accent. Very apologetically, he said he had been called away on business but would be with us as soon as he could. We folded our sheets and ropes then went to their canteen for a bite to eat while we waited for the American to return.

They certainly had things well organised. In the corner there was a microwave with a quick-service machine alongside it. You chose what food you wanted, pressed the appropriate button and put the delivered meal into the microwave. It was all paid for by the company.

Chuckles and I filled our plates with steaks, chips and peas. We took a couple of beers from their fridge and the meal went down a treat. Chuckles turned to me and said, 'They won't miss a few steaks. I'm going to put a couple in a bag for the wife.'

We sat in my cab and ate so that no one would know we'd been in the mess room. Unfortunately, we lost a considerable amount of valuable time; it was four hours before anyone turned up.

They didn't take long to unload our trucks. When they offered us something to eat we didn't turn it down, of course. It had been four hours since we had last eaten.

Before we left, we were approached by a rather portly man with a ruddy complexion, huge smile and an enormous cigar. He handed each of us a five-pound note, saying that it was for the delay. He hoped it hadn't put us out too much. A fiver was a lot of money in those days, so it hadn't turned out such a bad day after all. Free grub and a fiver – enough to satisfy anyone.

I followed Chuckles out of the gate and we were soon making our trucks sing as we drove down the A12 in the dark and rain. On some of the narrower stretches of road the driving was a bit hairy at times and by the time we were back past Ipswich we were over our driving hours. Chuckles still wanted to get home but I was praying that we didn't get stopped and checked. If a company truck driver got caught driving over his hours he would get immediate dismissal. But I knew that if I, as an owner-driver, were caught I would lose my operator's licence.

On our approach to Witham Chuckles indicated and pulled into a lay-by. I followed, wondering why he had stopped until I spotted the smashed-up AEC we had passed in the morning. I noticed that the trailer had now been removed.

We jumped down from our cabs, and as we stood gazing at the wreck Chuckles remarked, 'It definitely looks like an insurance job to me. They won't miss a few parts. I'm going to take the fuel pump off first.'

'That's a bit dodgy, isn't it?'

'Nah. Leave it to me.'

'Look, Chuckles, why don't we park your truck on the next lay-by and put the curtain round it so it looks as though the driver is having a kip. I'll drive you back here in the Volvo, then you can take what you like off and no one will be any the wiser. When it becomes known that there are a few bits missing from it, they won't suspect the driver of a brand-new Volvo,' I said, with a wink.

'You know what, Rich? You're more of a crook than I am.'

Chuckles threw his toolbox down by the passenger seat, then we drove about three miles to the next lay-by where we parked Chuckles's AEC.

I drove the Volvo around Witham and back to the accident lay-by. I drew the curtains. 'I don't know about you but I'm feeling a bit tired.'

'Me too,' Chuckles replied. 'Let's have an hour's kip.'

When we woke it was quiet and very dark but the rain had stopped. I took the bulb out of the ceiling light so if we had to run back to the truck the light wouldn't come on when we opened the door. We really did feel like a couple of crooks.

The door to the AEC was jammed but eventually we got it open. 'How are you going to see, Chuckles?' I asked.

'An engine is like a woman. You have to feel your way around,' he said with that saucy grin of his.

While I kept an eye out for the wooden tops, Chuckles eventually managed to get the fuel pump off. He handed it to me; it was very heavy. I undid the rope that tied my sheet down, pulled the sheet back and slid the pump under.

Chuckles carried on with his 'work', and after a while I became a little agitated. 'Are you all right, Chuckles?' I asked. 'We don't want the wooden tops stopping and asking questions.'

'All right, all right, keep your wig on. I'm just taking the head off.'

'What?'

'You heard.'

Between us we carried the spares back to my trailer. Some of the items were really heavy. Chuckles had taken the starter motor and alternator off and completely stripped the top end of the engine. There wasn't much left of the AEC after he had finished with it.

Afterwards we got rid of most of the grime on our hands with an old piece of diesel rag. It didn't smell too good. It was a good job neither of us liked cigarettes, otherwise we would have gone up in a cloud of smoke.

I was so relieved when I pulled off the lay-by. Chuckles wasn't at all worried – nothing seemed to bother him. When we reached the other lay-by Chuckles pulled the curtains from around his windscreen and started her up. The engine burst into life with a thunderous roar.

I glanced at the clock in the tachograph. It was quarter to three in the morning. Before long we reached the farm and unloaded the spares. We slept there for a few hours.

The three of us had now been working together for a year. Chuckles's truck had been stripped down and steam cleaned and it was sporting the new fuel pump which meant no more black smoke. We nicknamed her the 'Barnsley Special'. Benny had made a smashing job of painting our vehicles so that they all looked the same.

We had opened an account for our diesel with an

Esso garage on the A2. At the end of each month we pooled our resources and paid the bill. For tax purposes Jean was now our full-time secretary. She booked Len's old Land Rover as her transport and the diesel was charged to North Kent. I had soon realised that the fiddling was done at the top of the ladder and not the bottom.

There was plenty of work all over the country although the rates were not that good. However, we didn't grumble. At least there was work out there. In fact, life was sweet. We were making a reasonable living, the camaraderie between us was excellent and there were no real problems. It all seemed worth while.

However, in the late 1970s the trucking industry was changing rapidly. In particular, the new trucks with sleeper cabs were making digs a part of road transport history. Among the digs that I knew of closing were the Lantern on the A40, as well as the Towers at Thatcham, Biggleswade, Basingstoke and South Mimms. Then there were the Black Horse just outside Exeter and Bob's of Coventry.

It was not just the sleeping accommodation that was closing – the cafés went with them. The night trunkers were now changing over in motorway services and eating pre-cooked meals alongside the modern truck driver.

I regretted some of this change. In the old days the transport driver had wholesome food cooked by the landlady. The roadside cafés used to offer a friendly service with good food at all times of the day.

The AEC Mandator had become obsolete almost overnight. Although it was an excellent truck it was nearly impossible to sleep in because the bonnet inside the cab was extremely high.

When we were offered a long haul that meant staying overnight, Benny or I would do it. Benny had made a board on which he could sleep when he needed to. He would really ache when he woke.

While we were servicing our vehicles one Sunday morning, Chuckles said suddenly, 'I want to buy a new truck – a Scania or Volvo.'

Benny and I looked at each other aghast. We did our utmost to talk him out of it because although there was still plenty of work, the rates were abominable.

Chuckles had a family to think about and we didn't want him to do anything foolish to put them into jeopardy. Although his truck had become something of a dinosaur, it didn't owe him anything. It was still very reliable. However, Chuckles remained adamant that he wanted a change.

On Monday morning the three of us loaded out of Blue Hawk, Erith, for Penrith. The rates for this job were excellent but unfortunately poor old Chuckles would have to sleep in his extremely uncomfortable Mandator.

Leaving Erith at five-thirty in the morning, we drove through the Blackwall tunnel, up through Archway and onto the M1. The North Kent logo looked wonderful even though the three trucks were different makes.

I glanced in my mirror and saw Benny pulling out to overtake me. With its Cummins engine, Benny's ERF could do seventy plus and he went past me like a bat out of hell. However, he was loaded with twenty tonnes of plasterboard, so I just prayed that he didn't have to do an emergency stop.

The Cummins engine proved to be one of the greatest engines of all. They were very reliable and in America would do a million miles in their lifetime. They were extremely heavy, which made steering hard, but they had so much power that they replaced the Gardner. Cummins's representatives would tell drivers to give their engines full throttle as this was what they were made for. Benny didn't seem to worry that they used a lot of juice.

After passing my Volvo (top speed sixty-five mph) the Flying Scotsman overtook the Barnsley Special which was capable of doing just fifty-three miles an hour.

Then, just before the Luton turn-off, in the distance I saw something shoot about twenty feet into the air. There was dust and carnage everywhere. 'Oh no,' I thought. 'Something's happened to Ben.'

When I got to the scene there were bits of rubber all over the road and Benny had pulled onto the hard shoulder. Chuckles and I pulled up behind him. Seeing that Ben was all right I gave him a piece of my mind. 'What the bloody hell were you doing travelling at that speed? The tyres you have on are not made for that. If you had Michelins like you first had on your trailer, no problem. But you can't drive at seventy on cheap tyres.'

Chuckles put his arm around him and told him I was right.

While Chuckles cracked the wheel nuts, Benny jacked the axle up and I got the spare wheel from the carrier. The three of us worked fast to change the wheel because cars were zig-zagging around the rubber on the motorway and we were in a vulnerable position. It became so dangerous that Chuckles rang the police from one of the emergency telephone boxes to tell them about the hazard. We couldn't get away from the hard shoulder quickly enough. Within fifteen minutes we had pulled in at Toddington services, about five or six miles farther on. Chuckles got the bar out of his cab then went around and tightened up the wheel nuts on the truck. We cleaned up and went for a bite to eat.

As we sat there Chuckles and I teased Benny, saying that he would be going to Penrith and back for nothing as he would have to purchase a new tyre. He hated being teased so he went off muttering under his breath in Glaswegian. Chuckles had to turn away laughing. I could see his shoulders going up and down.

As the three of us strolled across the truck park, Chuckles suggested that we made straight for Penrith and tried to get unloaded without worrying about a back-load.

Chuckles and Benny gave me the thumbs up, then we drove out of the park and onto the M1. We kept down to Chuckles's speed – it was going to be a long haul up the M6 to Penrith.

As we passed Knutsford I noticed that the sky was filling with dark clouds. I continued driving, humming to myself as I drove, then all of a sudden the rain came down in torrents. I could hardly see out of the wind-screen, let alone spot Benny and Chuckles.

As the rain eased a little and the sky grew lighter I saw Benny in the mirror but not Chuckles. I eased off the gas pedal and drove at forty-five for about four miles before indicating and pulling over onto the hard shoulder. Benny pulled up behind, climbed down from his cab and walked up to me.

'No Chuckles, then,' he exclaimed.

'No,' I answered. 'We'll stay here a few minutes and if the wooden tops pull up and ask why we've stopped on the hard shoulder, we'll tell them that we've just changed the blow-out you had on your trailer.'

We agreed. Then in the distance we saw Chuckles. We were really relieved when he pulled up behind us. He climbed down from his cab, soaked to the skin and blaspheming.

'What happened to you?' I asked

'Don't ask me, Rich.' And then he proceeded to tell us.

'I was driving along quite happily when I noticed a peculiar smell. As I drove up the M6 a truck driver passed me tooting his horn and pointing to the back of my cab. When I pulled over I found that the number six injection pipe had broken. It was still raining heavily. I got drenched as you can see. So, I got the hammer out of my toolbox and flattened the end of the pipe in order to stop the diesel pumping out all over the engine. So now the old girl's running on five cylinders.'

'You can't drive all the way to Penrith with your clothes wet through, Chuckles. I've got a shirt and a change of trousers in the cab, you can borrow those.'

'It would make sense if you or Benny got a move on,' said Chuckles when he came back dry. 'It'll save a lot of time if one of you gets going to Penrith and tips.'

We all agreed. I looked at Benny. 'You're the Flying Scotsman, you get going.'

Benny climbed up into his rig, hit the starter button and the Cummins engine crackled into life. When he pulled out from behind me, we waved to him as he barked up the M6.

'What are we going to do about this injection pipe then, Chuckles?'

'What are we going to do?' he repeated, laughing all over his face. 'I'll push it all the way to Penrith on five cylinders. Come on, Rich, let's burn rubber.'

As I followed Chuckles up the M6 I thought to myself, 'it doesn't matter what life throws at Chuckles he still laughs and pushes hardships away. He's a good man.'

Despite the twenty tonnes of plasterboard she was straining to carry, the remarkable Barnsley Special kept firing on her five cylinders. However, I had been behind Chuckles for at least fifteen miles and I was becoming very frustrated. It was certainly going to be an arduous journey to Penrith and we still had to go up the long hill past Shap. Although we would be driving on the M6, which was a new road, it was a long old drag for the AEC Mandator which would probably be brought to its knees. Still, knowing Chuckles, he'd have a rope and a strong chain, and if the worst came to the worst I'd tow him.

Eventually we were passing Lancaster and heading up to Tebay. 'Thank goodness,' I thought. 'We're making progress.' It seemed like an eternity when we finally arrived at Tebay.

Now we had the final stretch on the edge of the fells to do, to the east of Shap. The AEC was crawling at no more than eight miles per hour so I pulled alongside

Chuckles and held my rope up in the air. When he gave me the thumbs down, I drove on ahead again, willing his truck to keep plodding on. 'You can do it, Chuckles. You can do it.'

Eventually we arrived at Junction 40, the Penrith turn-off. I'd always liked the AEC but my respect for it now was enormous considering what it had just done with its 11.3 engine. I was thrilled to think that we had made it without any problems.

After leaving the turn-off at Penrith we had only gone about a mile when we saw Benny approaching on his way back after tipping his load. He flashed his lights in recognition and as the road was clear we stopped. Benny walked across to greet us.

'You've done well, Ben!'

'Yeah, it's a good drop with a forklift unload,' he replied. 'The Barnsley Special made Shap!'

'But only by the skin of its teeth,' said Chuckles.

Ben continued, 'I think the best thing we can do, Rich, is that I drop my trailer and swap over with Chuckles, and then he can make his way back home.'

Chuckles was highly delighted with that idea. He told us that he was definitely thinking of buying a new rig. Benny and I looked at each other with disapproval, only because we didn't want Chuckles to get into financial difficulties.

It didn't take the three of us long to swap the trailers over. Chuckles headed south while I followed Benny through Penrith to a large building estate. It was getting late but we soon had the sheets off. At long last the forklift was unloading us.

Ben and I felt really stressed, hungry and completely

worn out. It had been a very eventful day, what with Benny's blow-out and Chuckles's injection pipe breaking. We drove on down the M6 and pulled into the next services where we sat and ate a hearty meal, the first time that I had enjoyed a pre-cooked meal in ages.

By the time we finished it was six p.m. We decided to punch on until we had had enough. Mile after mile we kept driving south. 'That's another hurdle over,' I thought, as we passed Manchester.

After a while I started to feel really exhausted, so I indicated and pulled off into Watford Gap on the M1. Benny pulled alongside me on the lorry park. 'Thank goodness you've stopped,' he said. 'I've been nodding all the way down the M6.'

It was so nice sitting there drinking a mug of steaming hot tea and relaxing. We had driven way over our driving hours, but not having picked up a back-load it was the only way we could make transport pay.

We had a wash and brush up. By the time I climbed up into my bunk it was around half past ten, I didn't remember a thing until I awoke at five-thirty in the morning. I climbed down from my cab and gave Ben a knock. 'If we stop here for breakfast,' I said, 'we'll catch all the traffic at Luton. I don't know about you but I want to be in front of the traffic not behind. Let's drive until we reach the Merry Chest on the A2. It should be plain sailing from there.'

Two hours and fifteen minutes later we were walking through the door of the café where the delicious aroma from the cooking immediately reached my nostrils. I hadn't realised until now just how hungry I was. The café was packed with the dying breed of transport drivers: lads who all had a lot of character and knew how to rope, sheet and find back-loads.

We ate a hearty breakfast and it went down a treat.

When Benny and I reached the old farm, Chuckles was up and had almost finished fitting a new injection pipe on his truck. He greeted us with a broad smile.

'Come on, let's go and have a cuppa. You've timed it just right. Jean's just put the kettle on.'

As we approached the garden, Chuckles's son, Colin, was mending a puncture on his bicycle. He looked up; he had the same smile as his father. 'He's a chip off the old block,' Benny remarked. Chuckles just smiled and answered, 'He's learning, he's learning.'

As we sat around the table Jean spoke about Chuckles changing his lorry. 'It's not up to the motorways,' she said, 'and we know how difficult it now is to manage a truck without sleeping accommodation.'

'But this is going to cost a lot of money.'

'Rich,' she replied, 'we are better off now than ever before. I know it means getting into debt for a while but Chuckles deserves a new truck. He shouldn't have to suffer any more punching that old AEC all over the country. It's hard work. We'll always get by somehow. If the worst comes to the worst, I know he could get a job in a garage as a fitter.'

'You can do one of two things: go for it or get out,' exclaimed Benny.

Chuckles chortled out loud, 'I'll go for it.'

'So will I,' said Ben.

I looked at him. 'What do you mean?'

'I'll buy a new truck, as well. Volvo's got a new 290

74

out and I've heard that they are good engines. If you buy two trucks you get a better deal.'

Chuckles rose from the table to ring the Volvo agents just outside Canterbury. After a lengthy conversation with one of the salesmen, Chuckles told us that he wasn't going to work that day as the Volvo representative was coming up to see him in the afternoon.

'Well, I'm not working either,' said Benny. 'I might as well have a word with him too. Perhaps we can strike a deal.'

'Oh no,' I replied. 'Not more haggling. After he has listened to Benny he'll probably commit suicide, poor sod.'

Chuckles rang around for more work but they were only offering eight-tonne drops, so I told Chuckles that I intended joining the others in a day off. 'While we're waiting we might as well wash the trucks down,' said Chuckles. 'They could do with it.'

When we sauntered down to the old shed we spotted Len loading cabbages into a very small trailer with only one axle. When he threw a bag of cabbages onto the trailer he then had to climb aboard to stack it properly. Len was working relentlessly. He told us that he'd cut all the cabbages off the field and filled about a hundred bags on his own, starting at two the afternoon before.

We looked at each other and decided to give him a hand. Although the trailer was only eight feet long the three of us managed – with great difficulty – to stack the hundred bags of cabbages onto it. They were piled so high they looked really unsteady. As Len pulled the trailer up the muddy pathway to the loading ramp it rocked to and fro precariously. Then the load was beginning to shift a little, before the top row suddenly collapsed, then another went and they fell onto the road, blocking it.

Of course Chuckles immediately saw the funny side of things, doubling up in hysterics. That started Benny and me off. We were laughing so much that even Len joined in. I know it's soul-destroying to see all your hard work spilling onto the road but if we hadn't laughed we would have cried. Then we had to clear the road and re-stack the cabbages on the trailer. So after the laughter all four of us put our backs into it.

When we left Len the three of us washed our trucks with brooms. As we worked a new eighteen-hundred Morris pulled up and out jumped a man in his early twenties. He walked over to where we were, introducing himself as Mike the Volvo rep.

Mike was a laid-back sort of person and not at all pushy. As we stood talking, he told us that the company he worked for mainly sold tractors and ploughs but had also started this new truck venture. The Volvos had been brought down from Scotland after being imported from Scandinavia.

'Just now as I drove in I noticed a Volvo parked up,' Mike said. 'Who does that one belong to?'

'Oh, its mine,' I replied.

'Volvos are selling really well,' Mike continued. 'But they still want to get more into the UK market.'

'So, what are you trying to tell us, Mike?' Benny asked with a broad grin.

'Well, the manufacturers are giving a very good deal at the moment, and it's a good opportunity to buy a first-rate vehicle. Correct me if I'm wrong,' he said, pointing to our old trucks. 'But do you want to part-exchange those?'

'That's right. But only for the right price,' I said.

Mike strolled over to where Benny's truck was parked. He took a notebook from his pocket and started to write down some particulars, then had a look at the other two. When he was satisfied, he came over to us again. 'I can offer you a good deal on three new Volvo 88s with the latest 290 engines.'

'Here it comes. I like a giggle,' Benny chirped up.

'Right oh. The deal is this. I'll give you four thousand for those three trucks and you can have the three new ones for eighteen thousand. After the trade-in that'll be fourteen thousand in all. How does that sound?'

We just stood there dumbfounded. As usual, Benny was the first to find his voice. He started to haggle unmercifully. The young man looked Benny straight in the eyes. 'I can see I'm wasting my time here!'

As Mike started to walk towards his car, Chuckles nudged me. 'Stop him Rich. Stop him,' he whispered.

'Hang on a minute, Mike,' I shouted. 'How can we do business if you sulk and go walking off?'

He looked at the three of us with an embarrassed expression. 'You can sit here all day arguing with me but I still can't give you a better deal than I've offered. I'll guarantee that within a year the price of these vehicles will be double what you've paid now. That is, if you buy them, of course.'

'I'll have one,' Chuckles replied. Then the two of us chorused, 'You had better make that three.'

'That's fantastic. You won't go wrong with these vehicles, that I can assure you.'

I asked Mike what sort of price he would get for the AEC when he sold it.

'Well,' he replied. 'I shouldn't be telling you this, but the company will get their money back on the Volvo 88, and the ERF will be a bonus. But they definitely won't bother about the AEC. It'll be farmed out to another dealer.'

'In that case, can we keep the AEC, Mike?'

'No problem. And if you agree, I can get the three vehicles sprayed for you – it'll take two weeks maximum. I'd like to enter them into the Kent county show at Detling as an advertisement for Volvo. The three brand-new Volvo trucks will be a sight to behold.'

'That sounds fine to me!' Chuckles exclaimed.

'But now, Gentlemen,' said Mike, 'to clinch this deal you unfortunately have to sign a few papers.'

Sitting with the papers up at the house it suddenly occurred to me that before today it hadn't entered my head to change my truck. It just happened. Talk about go with the flow. Things had progressed so fast.

We agreed unanimously that we wanted to keep the same colours.

Mike described how it would look. 'The North Kent logo will be on top of the cab in lights,' he said. 'The company will fit three external sun visors. The blind, painted in two colours – green with a red strip and shaped as a diamond – will be superb. It'll look really nice on the Volvo 88. It's the new Continental style.'

This was the icing on the cake and so we shook hands to close the deal.

'We've finally done it,' said Chuckles. 'We're in debt

up to our eyebrows, but what the hell, you only live once. Things can't get any worse. The UK still has to come out of recession.'

We all laughed. 'Bollocks to it.'

'But you surprised me when you said you wanted to keep the old AEC,' Chuckles said.

With a twinkle in my eyes I replied, 'Wouldn't it be a nice gift for Len?'

Chuckles slapped his knee. 'What a brilliant idea.'

'I've an even better idea,' said Benny. 'Our trailers are going to look obsolete with the new rigs. If we're going to spend all this money we might as well have the trailers refurbished and converted into taut liners.'

'Yes,' Chuckles agreed. 'We should have them made so that they can be dismantled at any time so we could use them as flatbeds. Ellis at Tonbridge are the best body builders around here.'

Ellis agreed to fit our trailers in between other jobs. After we had had seven days' rest – and to our complete surprise – Jean had a phone call to say the trailers were ready. When we went to Tonbridge to collect them we were dumbfounded. The blue canvases and red paint-work looked brilliant. They had also fitted removable side-boards.

The builders showed us how to operate the curtains and we were now the proud owners of three new trailers.

## Chapter 6

# Trucking with Less Stress

AFTER our break work started to pick up again and kept us busy, so before we knew it another week had passed. On a Wednesday night Jean told us that Mike had phoned to say that our trucks would be at the Kent Show from Friday to Monday and ready for the road on the Tuesday. This really lifted our spirits.

We left Gillingham in the afternoon in my car. Benny was up front, Chuckles and his family were in the back. Very soon I was driving into the county showground. It was a really spectacular event, displaying everything from tractors and machinery to all the produce grown by farmers in Kent.

Chuckles told Jean that we would meet up with her and the children later on. Benny was distracted by some of the other stalls but Chuckles and I couldn't wait to get to see our trucks.

Then as we walked I suddenly spotted the Volvo stand in the distance. We rushed over like little boys going after new toys. I stretched up to open one of the truck doors but found it to be locked. Then I saw Mike standing with some of his colleagues.

I asked him if all three trucks were the same. He said all three were 290s with sixteen-speed gearboxes.

'Mike, can you do me a favour?' I asked, winking at Chuckles.

'What kind of favour?'

'Well, when Benny catches up with us and we are all together there's a little story I want you to tell him. Watch his face when you say this and get ready to run.'

I told Mike what I wanted him to say. 'I'll go along with that,' he replied with a mischievous grin on his face.

Ben soon appeared from around the corner. Mike kept a straight face. 'A slight problem, Ben,' he said. 'They could only supply two trucks with a sixteen-speed gearbox. Rich and Chuckles have chosen theirs and the one left for you is the eight-speed. You won't mind that, will you? You're only eight short.'

'Eight short!' shouted Ben, his face getting redder and redder. 'Eight short!'

'Calm down. It's not the end of the world,' I said to him. We honestly thought he was going to explode, he was in such a rage.

'Calm down. Calm down! It's all right for you two, you're all right, Jack. Well, I'm not having it,' he said, looking straight at Mike, who was visibly shaking.

By this time Chuckles had had to walk off because he couldn't hold out any longer. His sides were splitting, and how he controlled himself for so long I don't know. I only just managed to keep a straight face myself.

Mike took his life into his hands when he said to Ben, 'Now who's sulking?'

Chuckles came back looking more composed. Then

the light suddenly dawned on Benny that we were winding him up and he had bitten it.

'You bastards. I really fell for that one, didn't I? Talk about hook, line and sinker. But don't you fret, I'll get my own back one way or another!' His face was still glowing.

On the Monday our trucks were due to be back at Volvo late in the afternoon, so Jean gave us a light lunch before we all made our way to Canterbury to collect them. Chuckles rode shotgun with me, and Benny followed close behind.

When we arrived they handed us the keys and necessary documentation for each vehicle. We couldn't wait to get behind the wheel.

As we walked from the office into the sunlight we were on cloud nine. The three trucks were parked up side by side in the yard. The sun was glistening on their cabs and they looked spectacular.

One by one we drove out of the yard. I savoured every moment of the journey back towards the farm and although I was used to driving my 88 this one was a different concept altogether. To me it was out of this world.

When we eventually arrived back, Benny and Chuckles watched as I backed onto my trailer. Benny was next, then last but not least Chuckles backed his on while we watched. We didn't want a major catastrophe with our new toys.

When the trucks were parked up the three of us stood back admiring them for a while. Benny and Chuckles jumped up into their respective cabs again, pulled their curtains and tried their beds. I just sat in my driving seat. 'I'm really glad I part-exchanged my 88,' I thought.

As I climbed back down I heard Jean calling to Chuckles, 'Don't you want any dinner today?'

I walked over to her. 'I'll tell him his dinner is ready and to hurry up otherwise I'll eat it for him.' She smiled and walked back towards the house.

The following morning I had a full load of sultanas to be loaded out of Denton Wharf, Gravesend, for delivery to Basingstoke. I decided to go on the A25 which would be a good test for the truck because it was notorious for its hills and it was a long old drag to Basingstoke. Normally I would have used back doubles through Bromley, Croydon and Mitcham.

I drove through Wrotham, Borough Green and then onto the A25 through to Reigate. The Volvo pulled like a Trojan. It was effortless to drive. I couldn't even hear the engine.

It didn't seem long before I was driving through Guildford and making my way to Farnham where I turned off and picked up the A287. Just as I was coming up to the A30 I noticed the storm clouds beginning to gather and that it was becoming very dark. I put my headlamps on; it was only one-thirty in the afternoon.

Then the heavens opened with hailstones and rain mixed, like nothing I had ever known in my life. I couldn't see properly out of the windscreen and imme-diately eased off the gas pedal, the traffic coming almost to a standstill. It rained so hard the soak-aways couldn't take the water and the road looked like a river. I just hoped my windscreen would hold.

I pulled up at the crossroads on the A30 and turned

left, punching south. As I started to descend the hill the rain was still coming down in torrents and didn't seem to be easing at all. When I put my foot on the exhaust brake the truck gave out a thunderous roar and I felt her ease back more.

Nature is a wonderful thing. As the sky grew lighter and the rain eased off I saw some beautiful wild flowers growing on the grass verge. They managed to survive no matter what the weather.

It was much brighter as I was pulling into the industrial estate on the edge of Basingstoke. Despite the weather I had enjoyed my first day out with the 290 – and the best part of this job was that it was a forklift unload.

I drove back the same way I had come and just before I got into Odiham I stopped at a transport café and had a mug of tea. No sooner had I drunk it than I was back inside the cab on my way towards the farm, feeling good to be driving a brand-new truck.

While I was washing the truck down Jean came out and told me that the three of us were loading sultanas again out of Denton Wharf for Basingstoke. That job lasted two weeks and was a good little earner.

In general we had a good bit of luck over this period and the three of us were bringing in plenty of money, so we weren't surprised when Jean gave us some good news one Sunday morning after we had serviced the vehicles.

She pulled open the drawer of the sideboard and retrieved a pile of papers, placing them in the middle of the table. 'You've done really well this last six months,' she said. 'All the diesel bills are right up to date. You'll also be pleased to know,' she continued, looking at each of us in turn, 'that one Volvo is completely paid off and we are well on the way to paying off another. The tax and insurances have been dealt with too. Considering the short time we've had the new trucks you've done extremely well. I must say this; you three have worked bloody hard and I'm proud of you.'

None of us knew what to say. Jean was such a lovely woman. Chuckles was a very lucky man.

'Oh, I forgot to mention there are two other things. Chuckles has found a nice little ten-tonne trailer. It's about twenty feet long and because the thirty-two tonners have made them obsolete the feller only wants three hundred pounds for it.'

'You're a dark horse, Chuckles. You didn't tell us,' I said.

'Oh I don't tell you everything,' Chuckles replied with that big grin of his.

'Well, what's it like?' Benny asked.

'It's in good nick. I've been thinking that as Len's been so good to us we could buy him the trailer and fit it on the AEC.' Without any hesitation Benny and I said it was an excellent idea.

'What's the second thing?' I asked.

'Well, the good news is you've got three thousand pounds between you for any emergencies that may arise.'

When Jean had collected her papers and we had finished our drink I offered to drop Ben off at the top of the lane. As he got into the passenger seat of my old car he turned and said, 'Perhaps we could have a short break.'

'That sounds great. We could all do with it,' I replied. Then I continued, 'Do you know something, Ben? Jean's done me proud.'

'Why's that, then?'

'Because it's my truck that's paid for.' I was trying desperately to keep a straight face.

'No it bloody well isn't!'

Needless to say, our journey to the end of the lane was silent. Ben had bitten once more.

For the next couple of weeks I was working for Associated Portland Cement delivering to builders' merchants all around the Kent coast. Most of the merchants' yards and loading bays were designed for six-tonne trucks, not forty-foot trailers, so manoeuvring was difficult.

Jean shared the work out between the three of us, alternating jobs to make it fair and square and not too tedious for any individual.

On the way to the farm one Sunday morning, I stopped off at a newsagent to have a look at the cards in their window to see if any properties were available. My eye was caught by a house to be rented in Railway Street, Gillingham. I didn't waste any time but walked hastily to the nearest telephone box and was very soon talking to the owner, Edmund Taylor.

Fifteen minutes later I met him in front of an old Victorian house. 'You've got references?' he asked.

'Sorry, no. I've done this on the spur of the moment after seeing your advert.'

While we were walking from room to room, I told him that I was in partnership with Charles Sorter. His face suddenly lit up with a broad smile. 'Old Sorter! We

86

were at school together. He's got a farm down lower Rainham Road.' I nodded and told him that I liked the house.

'Well, young man, I can see no problem but I'll have to confirm it with Mr Sorter. I've been let down before by tenants not paying their rent.'

We shook hands and I was soon on my way back to the farm.

As I pulled up to the old black shed, Benny was washing his truck. As soon as he saw me he turned the hose towards me in devilment and before I could close my window the water soaked me. I climbed out of my car shaking my wet arm. As quick as a flash I grabbed the other hose and doused him with it. We were drenched through, all in good fun.

When we had dried off and changed our clothes I told Ben about the house. 'I was thinking it would be ideal for the two of us,' I said. 'It's near the farm too.'

'How much is the rent?'

'He wants a hundred pounds a month, so that's fifty quid each.'

'It certainly sounds OK. If it looks all right we'll rent it,' Benny agreed.

While I was giving Benny a hand to wash his truck down, Chuckles walked over with a broad grin on his face, hands on his hips and his chest thrust out.

'My old mate Eddie has just phoned,' he said. 'After I educated him, he told me that the house was yours. Don't just stand there,' he continued. 'Get your arses round to the house quick. Eddie's waiting for you.'

We didn't need telling twice. We paid the deposit of two hundred pounds. I signed the documentation and

Eddie handed me the front- and back-door keys for the property. We all shook hands and he left.

Ben and I had another look around the house. The décor downstairs was liveable, but the bedrooms were covered in a patterned wallpaper which wasn't to our taste at all. We decided to strip the old wallpaper off and emulsion the walls. There were two single beds in the rooms. We moved these into the box room.

In the afternoon we drove into Gillingham to buy the paint. While we were there Benny said we might as well get the bed situation sorted out.

'Eh, Rich, we'll have to get double beds as you never know our luck. We might do some courting.'

'You're being a bit presumptuous, aren't you?' I replied.

'Well, I can always dream.'

When we told the sales assistant that we wanted two double beds to be delivered that day, she replied that it was impossible. Delivery would be in three days time.

'There's no such word as impossible. Anything's possible,' Benny jumped in.

'Well sir, they have to be ordered and it takes time.'

'Can we have the beds on the shop floor and collect them ourselves? We've got our own transport. Otherwise we'll take our business elsewhere.'

After conferring with the manager the saleswoman came back and told us that we could buy the beds but they would be sold as shop-soiled.

Benny immediately replied, 'Does that mean we get them cheaper?'

'Typical!' I thought.

'Well, yes sir. We'll discount them.'

'Right, let's get the necessary paperwork out of the way, and then we'll come back in about half an hour to collect them.'

We returned very soon, with me riding as a passenger in Ben's truck. The high street was very narrow, especially with an eight-foot-wide, forty-foot-long trailer but Benny managed to park quite near to the shop. By this time we were causing havoc with the local traffic.

When we walked in the door our two beds were on one side ready for us. As we loaded them on the back of the trailer with ease, the shop assistant said with a smile, 'I've never seen anything like this and I've been working here for quite a few years.'

By this time the traffic was really beginning to pile up. Impatient motorists were starting to blow their horns. I threw a rope around the beds and Benny tied the other side to secure them. We set off for the house.

I looked over at Benny. 'Isn't it great,' I said. 'We'll be sleeping in our own drum at last!'

Later that week the three of us had to be at Blue Hawk, Erith, by ten a.m. to load ceiling panels. These were to be delivered to Penryn in Cornwall. Although we were loaded by forklift, it was two o'clock by the time we tightened down the curtain-siders. In all it had taken us three and half hours. British Gypsum worked a twenty-four hour shift so as we had only done an hour's driving, we decided to truck as far as we could to the West Country.

Chuckles laughed and said, 'We are going to be night trunkers now.

The three of us agreed that we should go all the way to Penryn. Life was going to be a lot easier now that we all had sleeper cabs. Benny and I followed Chuckles to New Cross. We had a good run through as there wasn't too much traffic on the road to London and very soon we were driving over Battersea Bridge and along the Embankment. We went up past Earls Court and onto the M4 heading west. We stopped at the Heston Services for dinner.

As we sat eating our meal Chuckles said, 'When we go back to our rigs we'll remove the tachographs from the clocks and put in new ones in case we're stopped by wooden tops during the night. That'll give us another nine hours driving.'

We left Heston services at around four that afternoon. We climbed up into our respective cabs then gave each other the thumbs up signal for the off. We drove all the way down the M4 giving it full revs, known in our part of the world as the old Lord Harris, eventually ending up on the M5.

Personally I'd have rather gone onto the A303 as it wasn't as tedious as the motorway, but driving on the M4 we kept up a steady speed and in the long run it was quicker. In due course we were going around the Exeter by-pass then onto the A30. It was definitely stress-free to drive this truck compared to the others I had driven. A few years earlier the journey from Erith to Exeter would have taken me eleven hours and I would have fallen out of the cab when I got there. On top of which, the 290 engine in the Volvo was much quieter than I had experienced with the Gardner.

I didn't drop a cog until I got onto the A30 and the

only reason I changed down was because of the traffic in front of us. When we reached the other side of Launceston the traffic cleared and we made our Volvos tramp on again past the Jamaica Inn. We stopped in a lorry park at Bodmin because it was getting late and Chuckles thought it would be nice to get some fish and chips from a shop only a couple of hundred yards from where we were parked.

Chuckles was really pleased with his lorry. He said that although the journey had been a long one he didn't ache or feel exhausted like he normally did.

As we sat down in the fish shop Betsy the waitress came over to take our order. 'That was quick!' Benny exclaimed.

'Oh, I don't like to keep my customers waiting,' she beamed, showing off the dimples in her plump, round face. When she brought refills of tea she whispered to Benny, 'That's on the house.'

We teased him unmercifully on the way back to our trucks.

After the cosy warmth of the fish shop the air was cool outside. Once again we followed Chuckles out onto the main drag giving it plenty of wellie as we drove on down the A30, finally turning off onto the A39 heading by Truro to Penryn.

Chuckles eased back off the gas as we were now in a built-up area that was notorious for wooden tops. None of us wanted a ticket. We reached Penryn at about two-thirty a.m. It was so hilly that I put my foot on the exhaust brake but the thrust made enough noise to wake the dead so I removed it quickly. Instead, I did all the braking with the foot brake which in turn made the drums very hot. I wasn't a great lover of using the brakes as I had to look after the shoe leather. It was so expensive to fit new linings.

Chuckles pulled up alongside the kerb so Benny and I pulled up behind him. He had spotted a couple of police on patrol and walked over to them with his delivery notes in his hand. We watched as he chatted with them for a while. Then without further ado he jumped back up into his cab and we were on our way again.

We followed him through the town and then he turned off right. The lane we were driving down was so narrow there was barely enough room for our trucks. In the end we came to an enormous building around which there were caravans everywhere to house the building workers. As it was still early we drove just inside the entrance and had a kip. It was now three in the morning and very quiet.

I felt as though I'd only had a few hours sleep when I was woken by someone banging on my door. I gingerly drew back my curtains and wound the window down. Peering out I saw two larger-than-life fellows. 'Hello there,' they shouted up in Birmingham accents.

They were pleased that we had turned up early. 'Blue Hawk told us you wouldn't be here until around lunchtime,' they said. 'It was sheer luck that you were seen.'

'What's the time?' I asked.

'Oh, it's about seven o'clock.'

We were told to drive into an enormous complex which was all under cover and we noticed that a few trucks had already been unloaded. It didn't take us long to pull the side curtains round. Men seemed to be coming from every direction, straight from their beds, some still tucking their shirts into their trousers.

I had never worked so hard in my life as I did when I was pushing those ceiling partitions down to the men who were unloading me. They seemed to be having a race to see which could be the first gang to unload a truck. When the ganger signed our tickets, I said to him that I had never been unloaded so quickly. He told us that they were all on bonus and that they had to complete the ceiling in the complex within the next ten days.

'That's impossible, isn't it?' I said.

'Not when you've got the right blokes working for you. That's all they do – put up ceilings. Anyway, the sooner you're out of here the sooner we can lock up and go home for the weekend.'

We closed our curtain-siders, used their facilities and in no time at all were back on the road heading towards Truro. I glanced down at the tachograph. It was nine o'clock in the morning. 'Unbelievable!' I thought.

When we got back to Bodmin lorry park Chuckles rang Jean telling her that we would be home later that night and to find us work for the next day if she could. Then we walked into town for some breakfast. We were starving. After that we went back to our trucks and slept.

We left Bodmin at around midday. We gave it some stick all the way home.

This trip had been a good one. The rates were fine and we'd made it pay. It would have been nice if it could have been the same every day.

Jean told us she had had no luck in finding us work but would try again the next day. Ben and I were looking forward to returning to our own place – it was going to be great because we could do what we liked. Len and his wife probably felt the same about having their own space again now that Benny had left.

Benny walked into the front room of our Victorian house with two steaming mugs of tea. We agreed that it was really homely and comfortable. As we sat, there was a knock on the door. We looked at one another in astonishment. 'I wonder who that is!' I exclaimed. 'It might be a gorgeous, buxom wench.'

'I wouldn't be that lucky,' Benny shouted back as he went to open the door.

The woman who followed Ben in was a very cheery-faced, well-proportioned old dear who must have been seventy-five or so. She immediately thrust out her hand and introduced herself as Edie from next door. Ben offered her a mug of tea which she accepted gratefully.

Edie told us that her husband had died a few years ago and she still missed him terribly. They had lived there for more than thirty years, so she knew the entire goings-on in the road down to how many times the women changed their knickers. We could sense that she was lonely and wanted to mother us. So we weren't surprised by her dismay when we told her that we worked away a lot.

'I don't mind doing a bit of housework for you while you're away,' Edie said.

'You've got enough to do without doing ours as well,' Ben answered.

'I'd like to,' she replied.

So it was agreed that Edie would do the housework, and we would get a key cut for her. 'I could do some cooking for you when you're home,' Edie added. 'I'd like that.'

'That's kind of you to offer,' Ben replied. 'But it's only a short walk into town.' Then with a twinkle in his eye, he added, 'We can leave our dirty plates and cups to be washed by somebody else.'

'Oh I can see I'm going to have trouble with you two,' she giggled. Then she waved as she walked up the path to her front door. Ben and I decided to stroll into Gillingham to get a bite to eat. There were plenty of restaurants to choose from.

The first seven months in the new house passed so quickly. Edie turned out to be a real gem who couldn't have been a better neighbour. Many times she brought us in a home-made cake when we came back from a trip.

Road transport was another matter. Each day was becoming more of a nightmare. The three of us had been working day and night, snatching just a few hours sleep here and there. The only good thing was that we had now fully paid for our trucks. But despite that we were finding it hard to pay for our diesel. We managed only because – like a good many drivers – we put paraffin, kerosene and even red diesel in our tanks. We knew it was illegal and we were taking a chance.

Around the outskirts of London most trading estates with light industry were closing. The paper and steel industries were calling it a day, too. When there was work they expected drivers to carry their products for miserable rates.

In 1978 well-known transport companies with histories going back to the First World War were going into liquidation one after another, and with them went the old transport drivers who tried to hang onto their way of life. Callaghan's labour government had to go but worse was to come with Thatcher.

I don't know what we would have done without Len, because parking our trucks on his property saved us a lot of money. Chuckles's knowledge of repairing trucks was another great asset to us. If it wasn't for those two, I would have called it a day.

It came to the point where the three of us were confined to the farm doing odd jobs on our trucks because there wasn't much work around. We often turned down what work there was because the rates were so low that it wouldn't have been worth our while.

One morning when we arrived at the farm Jean said that the three of us were going to Manchester. We were to load rollers, gears and electric motors out of Northfleet, Gravesend, where they were dismantling the paper mills. This was very sad because over the years the mills had given good work to companies like Arnold's Transport and Hardy's of Northfleet.

The machinery was so bulky that we could only fit about ten tonnes on each vehicle. However, we were charging them as cap loads, which meant the full capacity of twenty tonnes per truck. Large companies like British Road Services never worked for less than eighteen tonnes cap.

We left Northfleet at half past ten and within no time

at all we were pulling in at the Merry Chest on the A2. We liked stopping there because it was one of the original cafés and we knew most of the drivers. After a hearty breakfast we had a good run through London in light traffic. Once we got onto the M1 we didn't change down until we pulled in at the Poplars, a well-known transport café on the A56 going into Manchester. There was no point in hurrying on because we knew we wouldn't get tipped that night.

We took a steady drive into Manchester and parked up at Trafford Park. The first pub that we chose was completely full with women's knickers pinned to the walls and ceilings. The landlord's fetish, we presumed.

After a few drinks we went into another pub, only to find that there was a coffin on the long table and that people were placing their glasses on it. People in Manchester certainly had a sense of humour. It made us all laugh – especially Chuckles.

The following morning we went to Blackley, about five miles from the city.

When we arrived we got talking to a self-employed driver from Blackburn whose truck was already being unloaded. Like us he was feeling the pinch. He said he knew of a clearing house in Trafford Park by the name of Johnson's and that they appeared to be getting a great deal of work

When eventually we were unloaded, we went to Trafford Park to make enquiries for a back-load to the south. We noticed quite a few trucks parked on some waste ground and that between two shops there was a clearing house. Inside, drivers waiting for work were sat down reading magazines.

We were stood talking to a driver from Grimsby when he suddenly turned towards the stairs. 'I know the bastard who's just walked up those stairs,' he exclaimed.

'Who is he?' Benny asked.

'The drivers have nicknamed him "Long Shanks". He's got transport clearing houses all over the country, but he quite regularly goes into liquidation and then moves on and starts somewhere else under a different name. A lot of drivers haven't been paid for the work they've done and there are quite a few truckers looking for him. I wouldn't like to be in his shoes when they catch up with him. As soon as I'm back in Grimsby I'm going to put the word out.'

'Good for you,' Benny answered.

'Well lads, I won't be taking any return loads from this place and I advise you three to do the same. You don't know how quickly you'll be paid, or even if you will be paid at all.'

While we were sitting in silence, thinking about what the Grimsby trucker had said, Long Shanks came back down the stairs. He was very tall and lanky with a ruddy complexion, long nose and thin lips. His suit looked shabby and in need of a good press. This man's features were implanting themselves firmly in my brain as he walked through the door and into the back of the office.

Chuckles looked at us and said, 'I don't know about you two, but the payment sounds too risky. Instead, we could be home in around five and a half hours.'

'Point taken,' we chorused. 'Let's go.'

As we walked out we spotted Long Shanks driving away in a brand-new green Jaguar.

We ran back down to the Poplars café and as we sat eating our meal we noticed that the Volvos and Scanias in the truck park were now being joined by DAF and Mercedes. This was the end of the old English vehicles like Atkinson and Foden which were three times heavier than any continental vehicle. They had been over-engineered with small power plants being made to labour with the weight of a train.

The continental manufacturers cut back on the size of the spring hangars and the chassis, fitting a more powerful engine into a lighter frame. They built the cabs higher which made the engine lower and then they put sound-proofing underneath the engine cover. They also fitted a turbo.

These continental machines were trucks for both the driver and the owner whereas the English ones were built for the governor only. The last thing they thought about was the driver's seat. They tried to mend their ways – but too late.

We decided to go home via the A5 because we knew of a garage on the old Watling Street where diesel was cheaper. As I cruised down the M1 at sixty miles an hour, all I could hear was the wind whistling and when I opened the window slightly it made my ears pop which I'd never had in any English vehicle because their engines were so noisy.

When I glanced in my mirror, Benny was right up close to my backside. If he were any closer he would have been in my driving seat. We pulled off onto the A5, passed the Hollies café and then punched on for the next five miles until we reached the Esso garage. Once on the forecourt we filled our tanks with about thirty gallons between the three of us, to get us back to Kent. Then we would diesel up again as it was a little cheaper on the A2.

As we left the pay kiosk, Chuckles remarked, 'I don't think my truck will pull so well now.'

'Why's that then, Chuckles?' I asked.

'Well, it's got used to running on red and kerosene.'

Benny and I cracked up.

When we pulled into the farm at four-thirty that afternoon. I told Ben and Chuckles that I was going to service my truck. Chuckles look bemused. 'Haven't you had enough for one day, Rich?'

'I've decided that I want Saturday and Sunday off. I feel as though I need a rest from work.'

'I'm all for that idea. I could do with a break too,' answered Ben.

'As I'm out-numbered, let's get cracking,' piped up Chuckles.

Chuckles adjusted the brakes on all three rigs, while I checked the engine oil and Benny looked for any air leaks. Then we walked up the road to Chuckles's house where Jean made us a mug of tea. Jean handed Benny and me three hundred pounds, and asked us if it was enough. 'Yes,' we said. 'Fine.'

She went on to say that everything was up to date and there was seven thousand pounds in the kitty, which was good. We were still holding our own.

'We'll need new tyres in three months,' Chuckles said.

'We'll have to keep our fingers crossed that it all works out OK,' replied Jean.

As Benny and I got up to leave, Jean said, 'Before you go listen to this. When Jenny was younger and people used to ask her name she used to say, "I'm Mr Sorter's daughter." I told her that was wrong. You must say, "My name is Jenny Sorter."

'I had completely forgotten about this until yesterday. I was down at the farm talking to Len when the vicar called round to collect his eggs. After we'd chatted for a while the vicar looked at Jenny and said, "You're Mr Sorter's daughter aren't you?"

'Jenny replied, "I thought I was, but Mummy said I'm not." Len and I didn't know where to put our faces.'

Benny and I laughed as we said goodnight.

As we got in the car Benny looked over to me and said, 'Let's emigrate to New Zealand!'

'You're having a laugh, aren't you, Ben?'

'No, Rich. I really want to go there. It's the nearest thing to Scotland.'

'That's a big decision to make!' I said. 'I'll have to think about that one.'

I started the car and drove out of the farm. All the way back to our house Benny talked non-stop about New Zealand.

The following morning I yawned as I clambered out of bed. As I passed Benny's room I shouted out, 'Get up you lazy bastard.'

'Hey you. Not so much of the lazy. Just get that kettle going, laddie. I've got a mouth like the bottom of a bird cage.'

I laughed as I went downstairs. Edie had put a cheerful vase of flowers in the front room. I made a pot of tea and then yelled up to Ben, 'I'm going into town to buy some new clothes. See you later.'

I'd never seen him move so fast. He came down the stairs so quickly he nearly went arse over tip. In the next instant we heard the key turn in the front door and Edie appeared.

'Morning, boys,' she said cheerfully, appearing not to notice that we were in our underpants. She poured herself a cup of tea and joined us at the table. We both stood up and said we'd better make ourselves decent.

She looked at us and said, 'Aren't you going to finish your tea first? Don't mind me. I've seen a man in underpants before, you know.'

We looked at each other and sat down again. We must have seemed like two naughty boys.

While we were finishing our tea we thanked Edie for keeping the place clean and tidy. She told us it was no trouble at all and that she enjoyed doing it. It gave her something to do. As we chatted, Edie told us that the window in Benny's bedroom gave her a different view of the street and she could see more of what was going on. We both cracked up.

'How could you be so nosy, Edie!' Benny exclaimed.

'Quite easily,' she answered, so cool, calm and collected.

As she walked to the front door she looked back over her shoulder and shouted with a gleam in her eyes, 'Have a good time but don't do anything I wouldn't do.'

After breakfast in a Gillingham café we took a bus to Chatham where we bought our new clothes. That evening, suited and booted, we went to a disco in Canterbury Street, Gillingham. The atmosphere was

102

pleasant. There were plenty of couples but also quite a few women without men.

'Hey Rich, look over there,' said Ben.

When I looked there were two good-looking women seated at a table. One of them got up and walked up to the bar where we stood. She ordered drinks for her and her friend. When she joined her friend again we walked over and spoke to them. They were really friendly and asked us if we would like to join them. We didn't need asking twice.

During the conversation we found out they were both divorced, and later on we all made it clear that we just wanted some fun with no strings attached. They came back to the house and the four of us had a bloody good weekend.

On the Monday morning Chuckles, Ben and I were back in the saddle heading north. I don't know how Jean did it but she had managed to get us work out of BP Aldridge near Walsall. There was a tight deadline. They wanted a hundred and sixty tonnes to be delivered to Sheerness dock in two days or under. Jean said that she would get Side Valve in to help out. He was an owner-driver and a good worker. When he smiled it was a half grin, hence the nickname of Side Valve.

In heavy Monday morning traffic we drove as quickly as possible up the M1 and M6, stopping at Corley services. Once back on the M6 we pulled off at Junction 7 around Walsall and into Aldridge where we drove automatically onto the weighbridge. BP were very strict and it had to be thirty-two tonnes dead or under, fully loaded, including the weight of the vehicle.

Definitely not a pound over or they wouldn't let the driver out of the gate.

When the forklift drivers started to load us they asked how we were going to work the tonnage. We told them there was another lorry arriving later and the four of us would be back the next day. Our trucks did look smart and fitted in well with BP's work.

Benny was loaded first so I told him to get going to try and get unloaded. We weren't going to have to stop on the way home because there was an excellent canteen and facilities on site, saving us a lot of time.

Chuckles went next and I followed afterwards. Meanwhile Side Valve drove in and I told him to try and get unloaded that day as there would be another load for him on Tuesday.

It was certainly a good turn around at BP so I was only three-quarters of an hour behind Chuckles. I gave it the 'Lord Harris' all the way back to Kent, down the M2 to the Isle of Sheppey.

As I drove into Sheerness dock the utility copper on the gate told me what shed I had to report to. They were a funny lot, full of piss and importance.

I was relieved to see that Benny had been unloaded; and that Chuckles only had three more pallets left on his truck. When the forklift driver looked up and saw me I could see right away that he wasn't a happy man.

The three of us soon had the ropes and sheets off. The men were only working until seven p.m. so when Side Valve eventually turned up he was much too late to be unloaded. 'Bloody lazy dockers!' I thought to myself.

Our emergency plan was to borrow a spare trailer from an owner-driver known as Wing Nut (large, bulbous ears) and run back up to Walsall early the following morning. We would fetch back four loads in good time to unload them as well as the trailer left over from today.

This depended on our being able to get hold of Wing Nut. At first Jean had no luck contacting him, so while we were all drinking a cup of tea at the farm we tried to think of other ways of working the tonnage from BP. We didn't want to hire a trailer – it would work out far too expensive for us.

Then at last the telephone rang. When Jean had finished speaking she told us that Wing Nut would be over as soon as he could and would bring the trailer. So Benny and I waited for him to arrive. It was music to our ears when we heard him finally drive in with his Scania, which sounded just like Chuckles's old AEC.

Chuckles checked the air hose couplings; they were the same as those on Side Valve's rig. We all mucked in and very soon had the trailer coupled up with the unit.

We thanked Wing Nut for the loan of his spare trailer. 'No problem,' he said. 'Chuckles has been very good to me in the past, sorting out different things on my trucks. And after all, there's one good thing about drivers, we all stick together through thick and thin, especially in a crisis.'

The following morning we met at four. It was great driving an empty truck through London with so little traffic.

Still before dawn, as I was cruising up the M1 at sixty, I saw in my driving mirror that Benny had pulled out and was trying to overtake me. He must have been in the middle lane for about five minutes.

Then all of a sudden there was a loud roar and the sound of horns blowing. My heart missed a beat. I might have guessed it was Side Valve making all the noise while undertaking me on the hard shoulder. The bastard had turned out all his lights so that I didn't spot him coming alongside me. Now I had Benny on my right in the middle lane, Side Valve on my left and Chuckles right up my rear. They were really playing games with me that morning.

After a while Side Valve pulled over in front of me and we kept Benny in the middle lane for about twenty miles. However, as we got up to the Blue Boar, the sky was beginning to lighten, so I eased off the gas and let Benny in. It was all in good fun. Fortunately there weren't any wooden tops around.

We pulled in at BP Duckham's Oil. While we were waiting for the forklift drivers Benny threw yet another wobbly, this time about Side Valve and I keeping him in the middle lane. Side Valve was such a placid sort of man that it was difficult to pick an argument with him. He politely told Benny to go back where he came from.

As this only made Benny worse, I whispered in his ear that if it weren't for Side Valve and Wing Nut we wouldn't have been able to complete this job. 'Go and apologise to him,' I said.

'You can get stuffed!' said Benny, stomping off. We just left him to calm down.

We all agreed that the first one to be loaded should get away sharp so that he could tip, drop his trailer and then pick up the other trailer and deliver it to Sheerness dock. Benny realised he had been out of order and

106

volunteered to do the extra load. It was a typical sort of apology.

We did well that week with first-class rates. For the rest of the week we were working out of Honix Wharf in Rochester.

## Chapter 7

# Guy Fawkes is Coming Earlier

IT was now autumn 1979. I had worked bloody hard but had little to show for it except that all the trucks were now paid for. Road transport was at its ebb and Jean was finding it more and more difficult to find us work. A few companies in North Kent were going into liquidation. The one thought that kept me going was, 'it can only get better'.

One morning Jean shouted out, 'Rich, I have a load for you out of Woolwich. Dalton Pipes for Derby.'

The day turned out to be catastrophic. It took all day to load. All I heard was, 'Load some of this.' 'Pull up.' 'Load some of that.' 'Move the truck up again.' 'Load some bends.' 'Move up again and load offsets.' This went on all day and by the evening I was pulling my hair out in frustration.

I was finally loaded by six o'clock that night. My number-one priority was to make the truck pay, so I decided to book on at Woolwich. I took my tachograph out, tore it up and put a new one in. Now I could night trunk it up to Derby and pull in at the Blue Boar for a bite to eat. I had all the time in the world.

At the truckstop I sat down where there were a few other truckers. They were saying how hard transport was becoming. Companies in Lancashire were calling it a day and you could now buy a good second-hand truck in any part of the country for a song.

The bantering and jokes were endless, most of them not suitable for family viewing. It really made me feel a lot better after having a good laugh with the lads.

After a meal I looked my rig over, climbed up into the driver's seat and checked my mirrors to make sure I had a clear view right down the unit and the forty-foot trailer. I made myself comfortable in the driving seat, turned the key under the steering wheel and she flew into life.

I was once more punching up the M1. Diesels seem to love being driven at night. They pull better and sound different. It must be the night air. I seemed to be driving effortlessly up to Derby. I thought that if I reached the site at around eleven thirty, I would be able to park up outside the job, get some sleep and still be the first one off-loaded in the morning.

When I arrived I asked a couple of lads who had just come out of the local pub where I could find Derbyshire Drainage. They put me straight onto it. I drove up to the locked gates so nobody could drive around me, took my tacho out, turned all my lights out, drew the curtains, stripped off and hit the sack.

A bang on my door soon woke me up. I felt as fresh as a daisy after having a good night's kip. I wound down the window and showed the fellow my delivery notes.

'Right oh, laddie. Put your draws on, then drive in and get your side sheets off.'

I had a very pleasant surprise because I had a lot of

help with the off-loading that morning. As soon as I had tipped I put my tacho back in and was driving back out by half past nine. I made my way to the nearest phone box and told Jean that I had spent the previous day loading and didn't want to go back there any more.

When Jean told me that there was no work at the moment I decided to ring the concrete works at Hulland Ward on the Ashbourne to Belper road in Derbyshire. Luckily they had twenty tonnes for Maidstone council.

Half an hour later I was driving onto their weighbridge, weighing myself and getting the tickets from the office. Out came the forklift drivers and in another half-hour I was fully laden with paving slabs. I roped the back end up inside the taut liner, drove back onto the weighbridge to weigh myself out then made my way to the first services on the M1.

I grabbed a quick sandwich, rang Jean and told her that I had managed to get a load. Then I punched down to Maidstone as fast as the old girl would go. I felt on top of the world. There was no wasted mileage and what had started off as a bad trip was ending up not so bad after all. If I could get this load off today I'd be onto a winner for the first time in a long while. And as for the tachograph – I couldn't care less.

When I got to the other side of Northampton the motorway traffic was building up, so it took me two hours to get to Luton. Then it suddenly cleared and I found that I was driving along at sixty again. It was hard to believe: one minute we were doing five miles per hour, the next sixty. I was trying to work out what caused the hold-up but I didn't see anything untoward or any obstructions. Ever since motorways were built these unexplained hold-ups seemed to have occurred regularly in all parts of the country.

'That's completely buggered my day up,' I thought. 'Maidstone will have to wait until tomorrow.'

Back at the farm Jean said, 'I didn't think you would get that load off today, Rich. But don't worry, we haven't any work anyway. I've had Side Valve and Wing Nut ringing me to see if I've got work for them. There's nothing. Ben and Chuckles are in the shed doing maintenance.'

I drove down to the shed. Chuckles asked me if everything was all right with my rig. I followed him as he started to walk round inspecting it. Although self-taught, he was a good mechanic who didn't miss a thing. After Chuckles had looked around, I said, 'Come on, Ben. Let's go and get some dinner.'

'OK lads, see you at eight tomorrow. Let's hope the work bucks up,' said Chuckles.

Benny and I drove straight home, parked the car outside the house and walked off into town. When we returned after the meal our old faithful was there to greet us. She never missed a trick. She made tea for the two of us then left us to our own devices. We put the TV on to watch the news but didn't take in a thing because we both dozed off in our chairs.

There was still no work the next day so Benny rode shotgun with me to Maidstone where we had a good turn-round in the council yard. Then we spent a couple of hours in the Rest café on the old A2.

Back at the farm we cleaned Len's AEC and trailer so that it gleamed when we had finished. It gave us something to do. It seemed strange to be driving one of the old English trucks again, and this truck had a unique sound.

Things suddenly woke up when Jean walked over to us and said that she had just learned that the ministry was checking Wing Nut's tachographs. 'Side Valve phoned me. I told him that you were here and he's coming straight over.'

'That doesn't sound too good,' said Chuckles. 'But at least Wing Nut hasn't had much work on.'

'Yes,' Ben agreed. 'He's had a lot of trouble with his truck so I shouldn't think he's had to run bent for any reason.'

'But if the ministry are in this area,' continued Chuckles. 'We've definitely got something to worry about because we've been running bent for ages. We've had to, to make it pay.'

In a short while Side Valve turned up. He was very pale and looked deeply concerned and agitated. I felt sorry for him. He was so worried he stammered as he spoke. He kept saying, 'I can't lose my operator's licence. It will finish me.'

Like the angel she was, Jean came out with cake and a pot of tea for all of us. As we sat, Chuckles suddenly said, 'Don't worry, lads. I've just had a thought.' His face was a picture, so I knew something was afoot.

Jean glanced at me with a frown and walked away. She knew Chuckles only too well.

'Rich,' he said. 'Drive your rig up to the house; we're going to put some office equipment on it. Side Valve, as far as anyone is concerned you have been operating from the farm. So you go home and get all your tachos and any old paperwork together, and bring it down here. Don't mention this to anyone, not even your wife.'

In the meantime we took my unit up to the house, cleared the office and loaded the contents onto the trailer. Jean, being a very wise woman, kept a low profile.

'We've got to work fast,' said Chuckles. 'The children will be home from school soon and I don't want them asking any questions.'

'What have you in mind, Chuckles?' I asked.

'Guy Fawkes is coming earlier than usual,' was his cryptic reply.

Then to Ben he said, 'Go down to the shed and fetch that worn-out old office chair.'

'OK, Rich. You and Benny take this lot down to your house but get back here pronto,' Chuckles instructed.

At the house we packed everything into the spare room, filling it so that the door could not be opened easily. Then I drove back to the farm where Side Valve was also back on the scene, his car laden with paperwork.

When we walked into Chuckles's house, Jean's office had been completely cleared of its furniture, pictures and other equipment. In their place were a worn-out chair and desk, an ancient typewriter, pictures that had seen better days, even a black telephone complete with dial. Chuckles had transformed the well-organised room into what looked like a hobo's den. 'I knew all that old junk in the attic would be put to good use one day,' he said. 'Now, don't just stand there, lads. There's work to be done!'

Side Valve told us that Chuckles had asked him to

sort out all his legal tachos and put them to one side. We sat on the floor and sorted out some legitimate tachos of our own, working quickly now because the children would soon be returning from school. As soon as we were satisfied with what we had done, Chuckles told us to wait outside in the garden, and as we were leaving we saw him light a paper torch and throw it into the office. It immediately caught fire and in no time at all the whole room was ablaze.

As Chuckles left the room he placed a shoe box of good tachos just outside the office door. He looked at us with a gleam in his eyes and said, 'I suppose I'd better ring the fire brigade, hadn't I?'

We just looked at him aghast. He was unbelievable. I could hear Jean screaming in the garden.

'Now don't forget, lads. When the firemen ask what happened we will say we were working down in the shed.'

While we were waiting for the fire brigade Len turned up on his tractor. 'Bloody hell,' he said. 'What's going on here?'

'The house has caught alight,' Ben replied.

'I can see that. Is everyone all right?'

'Yeah,' I said.

Len immediately took charge of the situation. He told Side Valve to turn on the outside tap. Then by mistake Len started to hose the bricks down outside until, to his embarrassment, he realised that the fire was actually in the building. He opened the front door and held the hose inside.

Within a few minutes three fire tenders turned up. Chuckles ran in to grab the shoe box of tachos. Len's face was ashen. I could see that he was really scared as he desperately tried to douse the fire.

As the firemen leapt down from their trucks, Len shouted to them, 'There's a man inside.' Just as they reached the front door Chuckles appeared carrying the smouldering box.

In fifteen minutes the Kent fire brigade had the fire contained and in thirty minutes it was completely safe to enter the house. When we re-entered, it looked like the Black Hole of Calcutta. The smell of burning debris was noxious. Everywhere I looked was wet and blackened with smoke.

After forty-five minutes the fire chief arrived in a small Bedford van. He was told by two of his firemen that they had been in the loft to investigate and had found that some wires had been chewed by mice. This might have caused an electrical fault.

It wasn't long before the local press arrived on the scene. The fire chief told them what he thought was the cause of the fire, and also mentioned that these old farmhouses weren't equipped like modern houses. They still had the old-fashioned fuse boxes.

When Chuckles had gone back into the building he had unknowingly burnt his hand. A rather large blister was beginning to form that looked worse than it really was. He went into the house and put on one of Jean's oven gloves wrapped in a layer of bandage to cover it. Then he came out again to face the press. When I saw him I had to turn away. I couldn't help laughing. I thought to myself, 'He's a better actor than Clark Gable. He's just got to get in on the act.'

Chuckles told everyone that his wife had alerted him

to the fire. He had immediately jumped into action, running into the burning building, just managing to retrieve the petty cash tin and some very important documents such as tachographs. He also mentioned how his brother-in-law had aimed the hosepipe at him to keep him wet while he was in there.

You had to hand it to Chuckles. He was something else. I'd never met anyone quite like him. I stood there listening to him making it up as he went along and thought, 'Any minute now he'll be telling them one of his jokes.'

But this was his best joke ever.

The fire chief praised Len for trying to cope with what was a very dangerous situation. He looked at Chuckles and said, 'You had better get that hand of yours seen to. You were a very lucky man. Take my advice and in future leave a fire to the professionals.'

'Yes Sir,' replied Chuckles. 'Yes Sir.'

They patted him on the shoulder as they left.

The local papers were full of what happened at the farm and they finished their story by saying that Mr Leonard Jarvis and Mr Charles Sorter were pillars of our society. 'We are very proud to have them in our community.'

While their house was being made habitable again, Chuckles, Jean and their children moved in with Len and his family. On paper Len charged them an extortionate rent, 'as the insurance company was paying'.

Within six weeks the builders had completely renovated the property so the Sorters were able to move back in. Jean was so pleased and proud of her house that she took Chuckles out for dinner.

As for poor old Wing Nut, he was reprimanded for sailing too close to the wind, and yet he was the only one of the five of us who had run legal. But we learned the lesson and – at least for a while – made sure that our tachos were legal too.

However, in the following weeks we grew increasingly concerned at how little work there was. Although the rigs were now paid for they still had to be taxed and insured.

One afternoon, after we'd been kicking our heels, Chuckles said, 'Now look, lads, it doesn't make any difference to me but if you feel that you can make a better go of it on your own, I don't mind at all.'

'We've been through a hell of a lot together. We're not going to break up now!' Benny exclaimed.

'That's great. But I don't want you to feel that you're beholden to me.'

'Let's hear no more of that sort of talk,' I said.

Benny and I left the farm that afternoon completely and utterly disgruntled with everything. We knew that eventually money would become a problem. Working for ourselves in road transport was certainly proving to be a gamble.

We were the proud owners of three Volvos which were all paid for. But there was no bloody work.

*Chapter 8*

# *Bless Her*

WHEN Benny and I came home one evening we noticed that Edie hadn't been around that day. There wasn't any milk in the fridge and that was most unlike her. We decided to have an early dinner in town then come home, get ourselves spruced up for the evening and paint the town red. I wondered if the two women we met the last time would be in the night club.

Before we left I jumped over the wall and gave Edie a knock. She took so long to answer I was getting worried, and when she finally opened the door I knew immediately that she wasn't at all well. Her face was ashen and gaunt, and her hair was dishevelled and lank. She did look frail standing there.

'Oh come in, Rich. I'm so sorry I've let you both down, but I felt terrible this morning when I woke up. So I made myself a cuppa and went back to bed.'

'How do you feel now?' I asked.

'Well I've still got a terrible headache and I just feel generally out of sorts,' she replied.

'Have you got everything you need? If not I'll pop to the shop for you.'

'No, No. I've got what I need. I'll be up and about tomorrow.'

As I turned to go I said, 'We certainly miss you. We can't manage without you, you know. You hurry up and get yourself better.'

After the meal Benny and I got ourselves down to the local night club. 'I'm feeling lucky tonight,' I said, 'how about you?' Ben just looked at me with a wicked grin all over his face.

No sooner had we walked up to the bar, than we were approached by two women. They were excellent dancers with terrific senses of humour so we had a jolly good evening with them. They'd downed quite a few gin and tonics which made them giggle.

We asked them if they would like to come back with us and have coffee, though we knew we didn't have any cream or milk. They didn't even notice.

I was glad when morning came and I could have a rest. When Benny appeared he looked like I felt, completely shattered. We both sat in silence as we drank our black coffee. We showered, shaved and then had breakfast in Gillingham. On the way back home we called in on Edie to check her progress.

Edie's milk was still on the doorstep which was a bad sign. So when I knocked on her door and there was still no response Benny and I became more than worried. Her neighbour from the other side came out to see what all the noise was about.

She told us that we would find a spare key under a flower pot by the back door.

I opened the front door and walked in, closely followed by Benny and the neighbour. I shouted out, 'It's only us, Edie.' There was no reply.

They followed me up the stairs and into Edie's

bedroom. The curtains were still closed and she was tucked up in bed. I shook her gently and said her name. Still no response, so I put my hand on her forehead. It was stone cold.

We stood there in shock for a while. Edie's neighbour was the first one to break the silence by saying that she would contact the doctor right away. She also rang Edie's daughter to break the bad news. Ben and I found it hard to believe that she had gone. We were going to miss her terribly. She was such a cheerful, inoffensive old soul.

Four days later Edie's funeral was held at the crematorium near the top of Bluebell Hill. Quite a few family and friends attended. After the service Edie's daughter came over and spoke to us. She said how much her Mum liked us, saying she used to call us her lads.

While we were talking she handed us each a small parcel wrapped in Christmas paper – although Christmas was still four months away – with our names on the tags. When we arrived home we opened our parcels. Inside each there was a pair of socks.

I looked at Benny and said, 'Bless her.' She was a lovely woman.

The year 1980 wasn't good for us. Although the August weather was beautiful, everything seemed to happen that month, what with Edie's passing, no work and then Benny starting to complain that he didn't feel too good. When I asked him what was wrong he told me it was nothing, so I gave up asking in the end.

Every day we congregated at the farm, sometimes joined by Side Valve and Wing Nut – I must say the five Scandinavian rigs did look immaculate. I would have preferred to see them all dusty and dirty because that would mean we had been out on the road earning money. But they were just standing idle.

We had the occasional fifteen-tonne load and every little helped because we had to keep the wheels rolling to keep our heads above water. We had a fortnight with no work. I spent so much time on the farm that I knew all Len's chickens by name.

Chuckles gave me a wink and said, 'If it gets any worse and we have trouble with any of the vehicles we will cannibalise Benny's truck.'

'I heard that!' Benny shouted. 'You two can both get lost.'

'Where's your sense of humour?' I shouted back.

One day Wing Nut told us that they wanted some trucks down at lower Shorne which was just outside Gravesend on the marshes, alongside the river Thames. 'They have built a new dockside and ships are unloading there,' he said.

'That's been kept under wraps,' Chuckles exclaimed.

'Well I'm off to see,' said Wing Nut. 'Side Valve and all the lads are already down there.'

It didn't take long for the four of us to climb up into our cabs. Our three trucks going along the road must have looked quite a sight. We turned right at the top of Strood Hill and made our way along the Gravesend road past Charles Dickens's House at Gad's Hill, down to Shorne cross-roads. Here we turned right onto what they now call John Mills's territory because Sir John made quite a few films along the Gravesend sea wall. Soon we were at Shorne Mead where the new dock was being built.

When we arrived we drove automatically onto the weighbridge. We gave our registrations and company names to the person in the security office, then we pulled our trucks over and reported to the transport office.

I told the woman in the office I could carry twenty tonnes comfortably. She typed my registration and company name on the delivery notes and handed them to me saying that the rates were fixed for Birmingham or Southampton. She added that if I were delayed for any reason I would be the loser, and the sooner I got the job done the better it would be for me. As I climbed up into my rig I thought to myself, 'This is going to be another carrot chaser.'

I made my way to a large shed where I could see an impressive stack of silver-coloured lead bars. The dock was very well equipped. The crane drivers soon had me loaded with bars while I stayed in my cab.

When they had finished I drove onto the weighbridge and everything was fine so I lifted up the side boards, pulled the side curtains round and adjusted the straps on the trailer. My load had to be delivered to the Cable works in Southampton. Benny, Wing Nut and Side Valve were going to Derby. Chuckles's load was for Pirelli's at Eastleigh near Southampton. We gave the thumbs up to each other and we were on our way once more.

It was half past eleven when I left. I really had to get a move-on because I had to be at Southampton before they closed. If I didn't make it, I would have to have a night out there which would make me late back the next day. I could lose out on getting another load and I didn't want that to happen.

The only advantage of leaving late was that I would miss the commuter traffic. It was foot down hard all the way and it wasn't long before I was driving around Mitcham Common and heading towards the Kingston by-pass. The load was very flat so I still had a good view all round, but when I braked and drove around round-abouts the old girl groaned and creaked. I suddenly remembered what Chuckles had told me once: if you were carrying a dead weight it was unlikely that it would move.

As I drove a feeling of contentment washed over me. It was great having work again. I stretched up and turned the radio on, something I couldn't have done a few years previously because back then the driver didn't have a radio in his cab. It was a pleasant run down to Guildford on the A3 where I picked up the A31 and travelled along the Hog's Back. I loved this part of England and this particular stretch of road.

I kept up the same momentum and very soon I was driving down the A30 which would eventually lead into Southampton. It had taken me four and a half hours which was really good going. When I arrived I was surprised to see lead stacked everywhere.

After the off-loading all I wanted to do was get back onto the A31 as soon as possible where I knew of a transport café on the side of the road. When I got there I put my tacho on break and relaxed with a mug of tea and a bacon sandwich. I was surprised to see Chuckles drive in because I had driven farther than him. I could only think that I had had a quicker turn-round on the job.

He joined me at the table and said how he hoped that the work would continue to flow in from the new dock. We hoped too that we would get paid on time because we were strapped for cash. 'We are a week's payment overdue at the garage,' said Chuckles with a worried expression on his face. 'I hope they let us derv up, otherwise we will be using Len's paraffin again.'

This new place at Shorne Mead didn't open until eight in the morning so I could see that the only way I could do the work efficiently was to make sure that I always loaded in the afternoon. 'As we are at present,' I said to Chuckles, 'we can't get loaded until the morning.'

'That's right,' he replied. 'The donkey can't keep up with the carrot.'

However, we managed to organise it so that we were collecting our loads in the afternoon and starting really early in the morning. If we left the lead overnight at the farm Len would keep the geese out to deter any unwelcome visitors.

Ben was still looking rather pale and wasn't his normal self. 'I've got problems down there,' he said and it was obviously bothering him.

'Best see a doctor,' I said – and to my astonishment he agreed.

The doctor said some cream would fix it but Ben was embarrassed and hated the examination.

On the following Sunday morning Chuckles said, 'Do you know what? I caught an old pigeon yesterday.'

Then Len piped up with a huge grin on his face: 'I caught a partridge.'

'Oh,' I said. 'I thought it was a canary.'

Then Jean poked her head around the door, 'Well,

what about poor old Benny,' she said. 'All he's caught is a thrush.'

I could see Benny's anger rising. All of a sudden he banged his fist on the table and stood up, his face as red as a beetroot and his eyes on stalks. He muttered something in Glaswegian and stormed out, slamming the door behind him.

It all went very quiet, then Jenny walked in and asked, 'What's wrong with Uncle Haggis?' We couldn't contain our amusement any longer and just howled with laughter.

When we had settled down again Jean said, 'We shouldn't tease him. You know what he's got is really painful.'

Like always it was quickly forgotten and we were soon back to normal. Work was coming in regularly now. The rates weren't the best but it was better than nothing at all. We managed to get extra loads which helped no end. We just did our best and got on with the work. Jean was receiving cheques regularly from the work we were doing.

The first thing that was looked at by wooden tops was tyres so we made sure ours were OK. All three rigs and trailers had had new Michelin tyres fitted which Chuckles would retread now and again with his small electric burner.

The months ahead looked rosy for us with plenty of work coming from Shorne Mead dock. Ships with different cargoes docked there from all over the world. However, we noticed that the drivers working out of this new dock were mostly owner-drivers. Looking back, we should have taken that as a warning

We were now loading 45-gallon drums containing some sort of solution. The turn-round was good so that the three of us were away by nine a.m. and heading for the Blue Boar services. We didn't dally long over breakfast as we just wanted to get unloaded and get the job done. With me leading and Chuckles and Benny following, we took a steady drive to north Wales. When we reached our destination it took about half an hour to unload each vehicle, which wasn't too bad.

It wasn't long before we were cruising along the M6 again. When we reached South Mimms our driving time was up. We had a good night there and as usual Chuckles made us and a few other drivers laugh. The bond between the three of us had grown very strong because we had stuck by each other through bad times and good. At times we teased each other unmercifully. Side Valve and Wing Nut were also good fun.

It took us roughly an hour and a half the following day to reach Shorne Mead dock where it was again time to load drums to go back to north Wales.

After a month of doing this we were going insane because there was no let-up. It was all work and no play – just boot hard down on the accelerator all the way to our destinations.

One day on the way home we by-passed South Mimms and drove back to the farm. We were completely dog-tired. Chuckles didn't look too good, and his sense of humour wasn't as sharp as usual. The three of us were exhausted. We had really worked hard.

When I got into my car I thought it wasn't going to start as I'd forgotten to put the battery on charge. It had gone completely out of my head. I had neglected the old girl a bit.

Ben and I sat in the curry house in Gillingham high street in silence; we didn't even have the energy to talk to each other. I paid the tab then drove us home. Once inside we sat in the arm-chairs and immediately fell asleep.

The next thing I knew it was seven o'clock, I gave Benny a kick, saying, 'Get a move on. We're late!'

'All right, all right. Keep your hair on.'

Before long we had showered and shaved, and once more I was driving us towards the farm. When we arrived Jean told us that she was having trouble getting poor old Chuckles up because he was still so tired.

Jean cooked us breakfast and while we ate she confirmed that the money had been coming in as regular as clockwork once a fortnight. The diesel was all paid off at the Esso garage and once again we had no debt at all.

'Now, do either of you want petty cash?' she asked.

'Yes please. We're brassic lint [skint].'

'This new company's got you over a barrel, hasn't it,' she said, handing us two hundred pounds each. 'They're taking the proverbial.'

The three of us knew Jean was right – but what else could we do?

As the weeks passed we were doing more and more, and I don't think there was any part of the country that we had not covered. I often wondered why the dock was in north Kent since none of the work was transported locally. Liverpool would have been much nearer to most of our drop-off points. Well, we just had to count our blessings.

The mileage we were doing was unbelievable. On an average day we were clocking up five hundred miles. By the end of each week we were on our knees despite all the advantages of our Volvos and sleeper cabs.

We got our heads together one night and all agreed that if we loaded late in the afternoon we could leave at two in the morning and punch on through the night. We would save a great deal of time and could get more miles in. It would also enable us to get back again to load the following afternoon ready for an early start the next day.

Coming back from north Wales I pulled into Knutsford Services where I spotted Wing Nut's truck. I found him sitting in the far corner of the restaurant fast asleep with his mouth wide open, his breakfast in front of him getting cold. I tapped him gently on the shoulder and woke him up. He immediately jumped up and grabbed an imaginary wheel. 'Hell, Rich,' he said when he had come to his senses, 'I thought I'd fallen asleep behind the wheel.'

While he was eating he said, 'Do you know what, Rich. I've not been home for two weeks. Side Valve has been away longer than that.'

After our meal we climbed up into our rigs. I was completed drained and must have slept for at least a couple of hours. I had a job to wake Wing Nut. When I did finally rouse him he wound his window down and told me his time would be up at Toddington Services. 'Where are you stopping tonight, Rich?'

'Oh, I'm driving back to the farm where I can shower, shave and put fresh clothes on.'

'What about your driving time?' he asked.

'Bollocks to the driving time. I've had enough, and if you conveniently forget to put your tacho in they can't have you for that.'

Wing Nut turned the key, opened the clock, took the tacho out and tore it up. He looked down at me with a grin. 'I think I've just forgotten to put my tacho in. Let's go home, Rich.'

We gave it the Lord Harris all the way down the M6 but as we approached the M1 Wing Nut started wandering. I drove up alongside him, blew my horn and shook my finger at him. When he put his thumb up, I waved him on and pulled back behind to follow him down to the Dartford tunnel.

We were certainly nearly killing ourselves by doing so much work but this meant that even though the rates were unfair our finances were working out quite well.

'Because you're driving a lot more miles we are spending more on spares,' said Jean as we sat at the kitchen table. 'But there's usually an average of three thousand pounds in the bank now which is one thousand for each of you. From petty cash at the moment I can let you have one hundred pounds a week.'

'Jean, there are not many men earning a hundred pounds per week just now,' said Ben. (It was 1981.) 'We are better off than working for a governor.'

Chuckles chipped in, 'Point taken. It's Saturday tomorrow. Let's have a day off. Unanimous agreement? OK, meeting adjourned.' Ben and I bade our farewells and made tracks for home.

For a change, Ben, who was a good cook, made us a meal, following which we retired to our beds.

Ben and I didn't wake up until about eleven. In the

afternoon we went into Gravesend and ambled through the Fort gardens breathing in the fresh air. The weather was glorious with a blue sky and the sun peeping out from white fluffy clouds. We left the gardens and watched the ships on the Thames and the smaller craft in the canal basin. It made a lovely change from driving all day.

Later that evening we saw a live show at the Woodville Hall theatre. It turned out to be the best thing we could have done. It made a really pleasant change for both of us.

I was woken the following morning by the aroma of bacon and eggs. It really made me feel hungry so I leapt out of bed, put my trousers on and was down those stairs like a dose of salts. Ben looked at me in astonishment. 'Where's the fire? If you think I've cooked you breakfast you're mistaken.'

I peered into the frying pan and saw four eggs frying nicely along with the bacon. 'I knew you wouldn't leave me out,' I said.

That Sunday we continued our holiday, starting with a visit to the Royal Engineers' museum in Brompton. For us it was like going down memory lane.

Then I took Benny to Rochester castle. 'It's magnificent. It's magnificent,' he kept saying.

Trying desperately to keep a straight face I said, 'In all seriousness though, Ben, they should knock this bloody lot down as we're short of hardcore.'

Benny went ballistic and started spouting off in Glaswegian. The only words I could understand were the expletives.

We saw some of the other sights of Rochester: the

cathedral where the sappers' names are carved in solid stone; the museum and corn exchange; and finally Restoration House, Dickens's inspiration for the house where Miss Haversham lived in *Great Expectations*.

'I've been thinking, Ben' I said. 'Perhaps we should look seriously into your idea of emigrating to New Zealand. We could go and do some research in London.'

'Would you go, Rich?'

With a twinkle in my eyes I said, 'I'm only going under duress and because you need someone to look after you. You're completely hopeless on your own and in any case you'd need me as an interpreter because they wouldn't understand Glaswegian.'

After that I gave Benny a little tour of the hop fields around Paddock Wood. It was another beautiful day. The skylarks were singing their little hearts out and everywhere was a different shade of green.

'I now know why they call Kent "The Garden of England",' Benny exclaimed. 'It really is beautiful here.'

That evening we spruced ourselves up again and made our way to the night club in Canterbury Street. There was plenty of talent there, no mistake.

'Oh no!' Ben suddenly exclaimed.

'What's up?' I asked.

'That woman approaching us is the one who gave me the parrot. I'm off to the gents.'

No sooner were the words out of Benny's mouth than he was gone. I felt a right chump standing there with a pint of beer in each hand. Suddenly a voice behind me said 'Oh, is that for me or your mate, Ben?'

131

I turned and there she was as bold as brass.

She looked me straight in the eyes and said in a soft, husky voice, 'Perhaps you and Ben can have some fun with me and my friend again tonight.'

I could have murdered Benny for leaving me to it, but I knew I'd have to tell her the truth or else we wouldn't be able to get out of it and it would spoil our evening. 'Here goes!' I thought.

'To be perfectly honest we won't be spending the evening with either of you tonight or any other night.' Then I told her exactly why.

'Oh!' she replied shocked. 'How does he know it's from me? It could have been anyone.'

'Well neither of us has had any other opportunity. We've been too busy working.'

When Ben finally reappeared I told him not to put me in that position again. He just laughed and said, 'It's my turn to get at you for a change, laddie.' There was no answer to that.

As we stood looking around the bar we were approached by three men who were speaking at the tops of their voices so that everybody in the bar could hear. 'You've upset one of our lady friends,' was the gist of what they said.

I did apologise but no sooner had I got my words out than the biggest of the three said, 'That's not good enough! Step outside.'

Benny looked at me. 'I'll go outside. When I shout "ready" you send them out one at a time,' he said.

I sipped my beer and stared at the three of them. 'Is that OK with you?'

'That's fine by us,' said the ringleader, 'and when we

have finished with this idiot we'll come back for you.'

Benny went berserk, nobody called him an idiot. His left hand shot out one way and his right fist the other. Within seconds Benny had grabbed the third man by his lapels and butted him. The fight was over in less than a minute. The crowd that had gathered around stepped back, anxious not to get involved.

Suddenly the bouncer came running in from the door shouting, 'What's going on here?'

I explained to him that the three men on the floor were looking for trouble, so they got what they deserved. He told us that the club was respectable and said that we had better leave.

The bouncer followed us, putting his hand on Benny's shoulder and starting to shove him along. This angered Benny who turned around and told the bouncer to keep his hands to himself. The bouncer continued to shove.

All of a sudden Benny lost it. He turned and punched the bouncer straight on the nose.

I heard somebody behind me say, 'Well, I've never seen anything like this. I thought I was going to have a good night out in the local club. Instead it's ended up like a pantomime.'

Another one said, 'Five minutes. That's all it took to lay out four men. Unbelievable.'

The manager said that he would let the incident pass this time but didn't want us to set foot in his club again otherwise he would call the police. When the other three men had received first-aid treatment they would be barred as well.

I had witnessed Ben's temper before but not quite as

bad as that. It must have been something to do with his coming from the Gorbals. They were a hard lot there.

As we walked down Canterbury Street, Benny said, 'I'm sorry Rich. I just saw red, that's all. I wish you had stopped me.'

'Not me. I didn't fancy spending the night in hospital,' I answered with a smile on my face. 'Well Ben, it was certainly a night to remember. We'll have to find ourselves another night club now, but when we do please keep your fists in your pockets, eh.'

He just looked at me and grinned. One of his eyes was a lovely shade of blue and black.

## Chapter 9

# *Truckers Stick Together*

THE work was continuing to come in thick and fast from Shorne Mead at rates which meant that to get a good living we had to work bloody hard. We really looked forward to our Sundays off.

In one of my worst weeks I left home early Monday morning to arrive at Chester by nine. Then I left at half past ten to make my way back. As I was passing Birmingham on the A5 I started to nod. I opened the windows, turned the radio up to full volume, switched the heater off and began to sing at the top of my voice. After a while my voice was so husky I had to stop singing. I even slapped myself around the face so hard that my cheeks burned. I was fighting to keep awake.

I eventually pulled in at Corley Services. I put my tacho on break and immediately fell fast asleep in the driving seat. I woke an hour later, and then had some refreshment. I climbed up into the cab, seated myself comfortably, touched the key and she flew into life. I looked through the passenger mirror to make sure everything was all right then looked through the driver's mirror.

I let the air hand–brake off, put the gear lever into

low third and the 88 pulled away like a dream. A car made more noise. I was heading south. My head was saying, 'you've got to get back to Kent and load ready for tomorrow.'

I was flat out doing sixty-two mph. I didn't ease up once. I felt the tiredness creeping up once more but relief washed over me as I drove through the Dartford tunnel with home finally in sight. I was hungry but I knew I couldn't afford to stop at the Merry Chest because I would probably have fallen asleep over dinner.

I pulled in at the Esso garage, jenked up with derv, removed my tacho and replaced it with a blank one and then drove on to Shorne Mead. They loaded me ready for Southampton the next day.

When I saw Len at the farm I asked him if he still kept his HGV up to date. When he answered yes I was really relieved.

'Why do you ask, Rich?'

'Can you do me a favour?'

'Of course.'

'Could you put your name on a blank tacho for me? Then I can say you drove from Gravesend down to Shorne and loaded it, then drove back to the farm.'

We walked back to the rig where I retrieved my tacho. I told Len where to put his name, then I put the mileage in and that made everything legal.

'I won't forget this, Len,' I said.

I went home and had not been in long when I heard the sound of air brakes. Benny had brought his unit to the house. He looked, as I felt, completely knackered.

We ordered a Chinese take-away and were in bed by

seven. As soon as my head hit the pillow I drifted off into a deep sleep.

At three in the morning I drove by car to the farm. It was a pitch-black night and there was no lighting so I made my way to the cab with the aid of a torch. I cruised slowly up the lane and away. I gave it the old Lord Harris all the way down to Pirelli's in Southampton – always a quick unload. In no time at all I was punching back up the A30.

As I was approaching the Hog's Back I pulled off into a transport café where the freshly baked bread always smelled delicious. After two more hours of monotonous driving I arrived back at Shorne Mead.

The transport routing clerk said there was another load to do for Southampton. I looked at him. 'You're having a laugh aren't you?'

'No,' he said. 'It has to be done.'

I was too tired to argue. 'They want blood,' I muttered under my breath.

I made my way around the back doubles of south London, driving on the outskirts of Croydon until I was on the Kingston bypass. Once more I was fighting sleep. I stopped in a lay-by for a jimmy riddle and walked around the truck a few times to get some fresh air into my lungs. Then I climbed back into the cab, lowered both the windows, checked the heater to make sure it was turned off, turned the radio up full blast, and made my way back to Southampton.

As I waited to be off-loaded, Benny and Chuckles pulled in. I climbed down from my cab and greeted them, and we stood chatting for a while. 'The rates are crap,' said Chuckles, 'but they have plenty of work. If

137

you can get two loads in you're earning money.'

'One load to Southampton should be enough for one day's work,' I argued.

Chuckles went on to say that he'd been to Dover, reloaded then driven back to Southampton and Ben had done the same. 'This working day and night is getting dangerous,' I said, and the others agreed.

Eventually we made our way back to the Hog's Back. We weren't sure what to do, go back through Guildford then pick up the old A25 and make our way back to Borough Green or drive up to the Kingston bypass and go around the back doubles eventually coming down the old A2.

'The A25 is a notorious road and you've got to drive through Reigate and all those stupid little villages,' Chuckles said.

'That's what I like about Glasgow. There are no villages in it,' Benny chipped in.

We finally agreed to go up to the Kingston bypass.

I hadn't driven far before I was nodding again. I noticed Chuckles wandering too, and he'd clipped the kerb once. Driving was beginning to be a bloody nightmare; fighting sleep was one of the hardest things we did in road transport. I could fully understand why it caused accidents. The old transport drivers wouldn't have carried on like this.

Eventually, after jenking up at Cobham, the three of us managed to get back to Shorne Mead and reload for Derby the next day. That day I had done roughly four hundred and twenty miles, loaded and tipped twice.

We all met up at three the following morning and left the farm at around half past. When we stopped at the Leicester Forest services for a quick bite Wing Nut and Side Valve were already there, having driven like the wind. None of us hung around as we wanted to get to Derby as quickly as was humanly possible.

Once we had tipped we had a leisurely breakfast in the Derby canteen, and chatted for a while. Once upon a time wherever I was I used to love driving home empty; now when I thought of driving back home I just felt dispirited. Working for Shorne Mead had completely knocked the pleasure out of driving.

Whoever was running the operation at Shorne Mead must have been making millions and yet the company was getting more and more greedy. We had come to the conclusion that an intermediary had obtained the contract and then had sub-contracted to owner-drivers. The intermediary took a large percentage which made the drivers' rates really pitiful. It was the same in other parts of the country at that time: individuals running clearing houses had no scruples. They made their hit, then ran.

When we arrived back at Kent and they asked us to do a load for Dover, Ben and I flatly refused. We were worn out. Side Valve and Wing Nut didn't like to say no because they were frightened that they wouldn't get called in again. How they carried on I don't know. All I can say is that they must have had more will-power than us. Ben and I loaded for the morning.

Back at the farm, and although he was as exhausted as the rest of us, Chuckles started to check the trucks to make sure they were up to scratch and roadworthy. 'We don't want to get pulled by the wooden tops through lack of maintenance. It's bad enough running bent.'

Len came over with a cheeky grin on his face. 'I'm completely shattered too. I've done at least three hundred miles driving HGVs.'

It suddenly dawned on me that they all knew that I had asked Len to fiddle a tachograph for me. The others had jumped on the bandwagon, so poor old Len was doing ghosters – booked down as driving when in fact he had not even turned the wheel.

Chuckles took the brakes up on the three rigs, and then we mucked in and washed the vehicles down with brooms. We agreed to start work the following morning at six.

Jean came out to greet us, her face pale and concerned. 'We haven't been paid for the last month. They are eight days overdue. They've never been this late in paying before. I'll give them a ring tomorrow to see what's happened.'

I told her not to worry too much and that I was sure there was a good reason for the delay.

The following morning, when Benny and I arrived at the farm, Chuckles was waiting and already had the three Volvos ticking over nicely. As we climbed up into our respective cabs Chuckles shouted over to us, 'First stop Leicester Forest.'

We drove out of the farm as quietly as possible then kept up a steady pace through the Medway towns and as we neared the top of Strood hill near Swain's Transport the A2 opened up before us. The three of us picked up momentum. We drove swiftly through London and just as we reached the top of Archway hill I thought to myself, 'We've cracked the first quarter of our journey.'

In no time at all the three of us made our way to Leicester Forest, keeping our feet hard on the floor for an average speed of sixty. We only stopped for a mug of tea, then we were back on the road to Derby.

On reaching Derby we had breakfast in their canteen while the men unloaded us. Chuckles seemed very reserved which was most unlike him. We kept fishing to try find out what was troubling him, but he kept changing the subject, so we left it at that for while.

We decided to stop at the Blue Boar on the way home and as soon as our rigs were empty it was foot down hard again and punching south. While we were sitting in the truckers' section of the Blue Boar, I decided to come out with it. 'Right, Chuckles. Are you going to tell us what's wrong? Is it because Ben's feet are chucking up?' As soon as I'd opened my mouth I knew I'd said the wrong thing by the look on Benny's face, but at least I'd managed to make Chuckles grin.

'I'm very concerned about the payment,' he said. 'The bank's crying out for their money and the fuel bill hasn't been paid.'

Again I told him not to worry. 'We'll get the money eventually.'

'It's all right saying that, Rich. We can go without food but we can't drive our trucks without diesel.'

When we filled our vehicles at the Esso garage we didn't have a red mark so up to now we were all right. We didn't know how long for.

At Shorne Mead there were about thirty trucks waiting to get loaded, all belonging to owner-drivers like us. It came to light that they were having the same trouble, not being paid for the work they had done.

Chuckles was getting on his soap box. 'We can't run trucks without any money,' he said. 'Here's what we do. We'll all get loaded, drive our trucks outside the gates and park them. We'll walk back inside and tell them we are not moving until we receive our money.' So that is precisely what we did.

Chuckles thought it would be best if two representatives went inside to talk to the management. All the drivers had great respect for Chuckles and agreed that he should be one spokesman. He wanted Benny to be the other one.

'Why Benny?' we asked.

'Well, if they won't take the quiet approach from me I'll let Benny loose on them. They won't know what's hit them if he starts.'

The loaders thought it was strange seeing all the trucks parked outside the gate while the drivers stood chatting amongst ourselves, some cracking jokes, but most becoming increasingly angry.

When Chuckles and Benny eventually came back from the office, we all gathered around them. 'At first one of the men in the office tried to fob us off,' said Chuckles. 'I wasn't having any of that so I demanded to see the manager, Mr Cruise. Eventually he called us in; we were cool, calm and collected, as they say. I explained to him that it was impossible to run our trucks without being paid and that some of the lads were in trouble with the bank.'

'Good for you,' chorused several drivers.

'Then Benny stepped in and told him that the drivers would not be moving their trucks unless they got paid. I could see the manager quaking in his boots.

142

'When Mr Cruise rang head office he asked us to wait in the other room, which was fair enough. One of the jumped-up traffic clerks told us that we had a cheek going in there demanding to see management and upsetting a well-organised company. I told him he couldn't organise a piss-up in a brewery.

'By this time I could feel my blood boiling and I knew Ben's temper was on a very short leash. However, Mr Cruise then emerged again. He told us that the bank had had a temporary problem processing payments. "You'll be receiving your money within the next two or three days," he said. "Every man deserves to be paid for the work he has done."'

All the lads seemed happy with what Chuckles told them so we walked to our trucks and moved off.

In the next few days we were running from Shorne Mead to Doncaster, driving there and back in one day. These were desperate days when we completely threw the law out of the window in order to survive. On one really cold morning I nearly froze through driving with the windows open in order to keep awake. I pulled onto the hard shoulder, ran around the truck to get the blood pumping then jumped back into my cab for another couple of hours before repeating the same thing all over again. I was frightened to stop to get some shut-eye because I would miss getting loaded and would earn nothing the next day.

Then two days after the meeting with Mr Cruise the woman behind the desk at the Cobham garage told me that we had a red mark by our name. 'You can't fill up,' she said.

'Look, we've been promised payment in a day or so,'

I replied. 'There's a temporary problem at the bank.'

'I know you well enough by now, Rich. Carry on. I haven't noticed any red mark.' (Nod nod, wink wink.)

I thanked her. Then as I climbed back into my cab Chuckles pulled in. I was just about to tell him about the fuel when he said, 'Rich, you know I was on locals today. Well, when I was going back to Shorne Mead for my second load who do you think I saw coming out?'

'Who?'

'Long Shanks with the green Jag.'

'You're joking.'

'No, Rich. I couldn't be more serious.'

'I don't like the sound of this,' I replied. 'I've smelt a rat for a long time now. I'm not going to do any more work for this crowd. I've been lucky to get this tank full of fuel and I'll use it carrying other people's products. Definitely not Shorne Mead's.'

'Well, two days are up now, Rich. I've been back home so I know we've not received any payment.'

It must have been our lucky day because the woman at the garage let Chuckles have fuel as well.

'I'm going to drive back out to Shorne Mead just in case they've got our money there. We've got to get paid. We've got to get paid,' Chuckles kept saying. 'Otherwise we are in shit street.' I'd never known Chuckles so stressed out like this. He was complaining of chest pains too.

I followed Chuckles back to the terminal where something was clearly amiss. 'We're not doing any more loads until we get paid,' I said to the routing clerk.

'There's no work for you to do, anyway,' he replied.

144

'The ships are late docking and won't be here until tomorrow. A lot of drivers have been having a moan at me about their wages but I've been told on good authority that your money will be here at ten in the morning.'

'Well at least that sounds encouraging,' I said. 'We'll be back tomorrow to collect.'

'One other thing,' he said. 'I've been told that none of you drivers can leave your rigs inside the depot because of the fire hazard. If you want to leave your vehicles here you must park them on the hard standing outside.'

'We won't be parking here. We're off home.'

As we drove out of the yard we could see some drivers already congregating on the hard standing. In the end there would be at least forty vehicles parked overnight.

That night Benny and I ate at Ernie's café in Trafalgar Road, Gillingham. I told Benny that Long Shanks had been seen at Shorne Mead but that our money would be there tomorrow morning by ten.

'Pigs might fly!' was Benny's response.

When he appeared at the kitchen door in the morning he looked like death warmed up. 'Ben, what you need is a bowl full of Scots oats to liven you up. On second thoughts a bloody good haggis would do you a power of good.'

He looked up at me. 'It's being so happy keeps you going isn't it, Rich. You bloody jumped-up little corporal. You've spoiled the lovely dream I had last night.'

'You don't have to tell me, Ben. It was about women.'

'No, laddie! It was about New Zealand.'

'If you think I'm going halfway round the world with a lazy, good for nothing moaner and, to top it all, a Scotsman – you're bloody mistaken.'

'They've got some large rigs out there, Rich.'

'I'll think about it. In the meantime let's go to the café for breakfast.'

'Can we afford it?'

'We'll say we left our money at home and put it on the tab. After all they know us.'

After sorting ourselves out and having a good breakfast on the slate, we drove down to the farm where we were greeted by Chuckles who was making jokes as usual. 'This bloke was trying to pinch Len's front gate,' he shouted.

'What did you say!' exclaimed Ben.

'I didn't say a word in case he took offence.'

We shook our heads and smiled. Driving to Shorne Mead wasn't very enjoyable that morning, as it was a miserable bleak day and drizzling. Across the Gravesend marshes it was so misty I could hardly see the cattle grazing in the fields.

When we arrived there was a lot of activity. Truckers were standing around in groups outside the gates and I guessed right away that something was wrong. We climbed down from our rigs and as we neared the gates the drivers stepped aside to give us clear access. I felt as though I were in a dream, everything was in slow motion. I didn't even question why they stepped aside.

It was only when I reached the gate that I saw the notice. My blood ran cold when I read the big bold

146

letters: ***CLOSED. GONE INTO LIQUIDATION.***

I was absolutely dumbfounded. I felt like dropping to my knees and crying. We had been running day and night and now we were in serious trouble with the bank. The three of us had a huge overdraft.

'Now I see why there were no ships in the dock,' said Wing Nut who had come over to join us. 'They must have known what was going to happen. It's all been done deliberately by top management. The only way out for me now is to sell my rig.'

We nodded vaguely but didn't respond, each of us lost in his own thoughts.

The other drivers were very restless and began to throw things at the gate in frustration. We heard Side Valve shout out, 'Let's pull the fence down with the trucks and burn the place to the ground.'

I glanced at Chuckles. 'We're in the shit,' I said, but he didn't seem to be listening. His face was as white as a sheet, his lips were quivering and his eyes were glazed over. He turned on his heels and walked away.

I felt sorry for Chuckles, he was older than the rest of us and had the added pressure of a family to support. I couldn't imagine how the three of us were going to survive this, there was just too much money involved.

When the staff started to arrive they were just as flabbergasted as us, because they too would be unemployed. On seeing one of the forklift drivers, Side Valve called to him, 'Geordie, you and your mates can lift up all the office Portakabins, run them along the pier and drop them straight into the Thames.'

Later, they did just that.

By now everybody was inside the gate, including the office staff. They all had strong reasons for behaving badly and nobody was going to blame them. There was a lot of noise going on, angry employees were breaking up everything in sight, nothing was missed. Even the penpushers were joining in and poor old Jim, the security man, kept saying, 'I haven't seen nuffink.'

We spotted the office staff putting office equipment such as typewriters into the boots of their cars. Who could blame them? It was all going to be broken or burnt anyway.

By now the forklift drivers were unsuccessfully trying to pierce the tanks of fuel with their forks. Ben told them to put their forks under the fuel pipes and lift the tanks up. That worked and everyone stood back as the fuel oil gushed out. No one thought about the consequences, we were much too concerned with getting even, and preoccupied with how we would survive this setback.

I moved through the chaos as if in a dream, I just couldn't believe what had happened. It was surreal.

Then all of a sudden Ginger Mitchell ran over, so breathless that he could hardly speak. 'Chuckles, there's something wrong with Chuckles,' he stammered. 'Shed eight, he needs an ambulance.'

Wing Nut went to phone for the ambulance while Benny and I followed Ginger to the shed. Chuckles was sitting where Ginger had left him. He was deathly pale and his skin was cold and clammy; he held his arms around his body defensively, as if in pain. 'Chuckles!' When I said his name he turned to look at me but his eyes were terribly frightened.

'Rich,' said Benny, hopping from one foot to the other in panic. 'They're trying to set fire to these buildings.' I realised that he was right, this was not a good place to be. We helped Chuckles to his feet and made our way to the main gate.

As we walked towards the gate the carnage was everywhere. The drivers had completely smashed the place up in their anger.

A driver from Dartford dipped a piece of rag in the spilt fuel then set light to it and threw it in. The explosion was deafening, and as the fuel was spilling out of the tank, flames were licking along the floor, setting fire to everything in their path. The devastation was unbelievable; but the satisfaction on everyone's face was a picture. They had settled the score.

The ambulance arrived almost as soon as we reached the gate and, thanks to Benny's mad waving, spotted us immediately. The paramedics put an oxygen mask on Chuckles's face and helped him into the ambulance. 'We'll call Jean,' I told him before they closed the door and then we watched in stunned silence as they drove off to the Gravesend hospital.

I called Jean from the only place on the site which had not been destroyed, Jim's security office by the front gate. I told her what had happened as gently as possible and although I could hear the shake in her voice, she was as efficient as usual: 'Len will drive me, I'll phone you at home when I know how he is.' I thanked her and hung up.

Jim had retired to the corner of his office with a newspaper and was studiously ignoring the goings on outside. He told me that he had to stay behind.

Whatever happened he still got paid a small but regular income for the two years until he retired.

'If anyone asks about this,' Jim continued, 'I'll say I was locked in here but managed to break out. "I didn't see nuffink," I'll tell them.'

'Would you do me a favour, Jim?' I asked. 'If you see a man who is very tall, slim and drives a green Jaguar, will you ring me at the farm?'

'I will if I can, of course I will. Mum's the word,' he said, touching his nose with his index finger.

By now the drivers had begun to climb back up into their rigs and as they drove off up the road in convoy, they left behind them a trail of devastation. I was anxious to follow them, to be home when Jean called. In the distance I could hear the fire engine sirens getting closer.

Wing Nut said he would drive Chuckles's truck home since he had ridden with Side Valve that morning. We left Side Valve with the promise to call him when we knew anything.

With only a brief detour past the farm to park up the trucks, the three of us made our way back to our little drum in Gillingham.

Once indoors I made us all coffee. 'I don't know about you two,' Wing Nut said, 'but I've had enough. I'm packing it all in. The old woman will go spare with no money coming in. But she'll be even more upset about Chuckles.'

We just sat there in silence listening to him rattling on.

'If I can sell my rig, I can square up all my debt and I'll still come out all right so long as I get a job right away. It's all right for Side Valve, he's single and still lives with his mother. I suppose you two will call it a day. . . .'

I looked over at Ben. 'Call it a day,' he said. 'You're joking, aren't you! After all we've been through, working our bollocks off. Chuckles and his family are in shit street and you think I'm going to call it a day. Never in a million years. I'm going to work day and night. I want to pay off my debt, and when I've got some money behind me I'm going to buy a house in New Zealand. And I'll tell you this, Wing Nut. Sell your rig to get yourself on your feet then come and work for us because we need you.'

'Need me! Why?' he asked.

'Because you're going to drive Chuckles's rig while he recovers. We'll pay you a driver's wages and the profit you make will help Jean and the kids until then. We can call it a day later on. North Kent is still going to be operational for a little longer.'

'How do you feel about that, Rich?' asked Wing Nut.

'I think Ben's got it all worked out. We'll tell Jean our plan when she's not so worried about Chuckles. So advertise your lorry for sale, Wing Nut, and tell the missus that you're going to work for us.'

'But how are you going to pay me?'

'Don't worry,' Benny piped up. 'Leave it to us. We'll sort your wages out.'

It was late afternoon before Jean phoned and we had long since fallen to watching the phone in tense silence. When it finally did ring we all leapt for it but I got there first. 'He'll be fine,' she told me, and the news was greeted with a unanimous sigh of relief. Jean explained that they thought that the heart attack had been caused by the stress of recent events and that we could visit the next day.

I laughed in relief, 'We should have known he'd be OK,' I said, 'It's Chuckles after all, and he does always bounce back laughing.'

## Chapter 10

# An Unexpected Gesture

WE visited Chuckles twice in the next three days and the second time we saw him he was looking much more like his old self. He even entertained us with a wealth of new jokes to show he had used the time productively.

When Chuckles came home we were at the farm to meet him and within five minutes were drinking tea at the kitchen table. He and Jean were quite taken aback when we told them that we would like to carry on with the business for at least two more years. 'Are you sure?' she said. 'You do realise how tight money is, what with the insurance, tax and MOTs, let alone the fuel. You need some petty cash too.'

It was then that Chuckles dropped his bombshell. 'I'm not going to be able to drive any more,' he said. 'In fact, I wanted to tell you – Jean has found me a job in Birmingham as a truck fitter, I've had it with transport.'

I was surprised, I had never imagined Chuckles doing anything other than driving his truck but I couldn't deny that he would be great at the new job. 'Wing Nut was going to do some work for us in your truck while you recovered. How do you feel about him doing that permanently?' I asked.

'Why not? The truck belongs to North Kent and it's no use stuck in the yard.'

We talked around it. Jean said she liked working in the office and Ben told them about our plans to emigrate to New Zealand once we'd saved some cash.

'Well let's hope everything works out for us all,' Jean said. 'But you know the old saying: act in haste repent at leisure. None of us can afford to be too hasty. So let's sleep on it first before we decide what's best.'

We were keen not to make any more mistakes after our recent experiences so took Jean's advice. The next couple of days were spent working on our vehicles. It was amazing what we found to do.

It was on one of these days as we were loafing around down at the shed that Jean called me to the phone. 'It's Jim,' went the voice at the other end. 'Do you remember me?'

'Yes. I thought I recognised your voice. You're the security officer.'

'Remember I haven't said nuffink. Right?'

'OK, Jim. Mum's the word.'

'Well, that tall, thin fellow you were on about with his Jag has just driven in. He's so angry he looks as if he's about to bust a gut. He asked me if I saw anything or rang anybody about all the devastation. I told him that I tried but a couple of big chaps threw me in the shed and locked me in. Now he's walking around the dock with his fists clenched as though he's going to punch someone's lights out. I'm hanging up now. Remember, I don't know nuffink!'

I shouted out to Jean that I would see her later as

something had come up. I ran down to the shed to get my car.

'What's going on?' Ben yelled.

'Long Shanks is at Shorne Mead!'

Benny jumped into the passenger side and we sped off up the lane. 'I hope that bastard's still there,' Benny was muttering.

In no time at all we were turning right at the top of Strood Hill and speeding along the Gravesend road to Shorne crossroads, the car leaning over as I turned right and the tyres squealing. Benny was hanging on like grim death. When I drove down the lane toward the Gravesend marshes I spotted the ships going up the river Thames. As we got closer I very nearly took off when I drove over the railway crossing. Benny's face was ashen.

In the distance we saw the Jaguar. 'Great! The bastard's still here,' Benny shouted.

I screeched to a halt as I drove in front of the Jag, completely blocking it in. As we walked through the gates we acknowledged Jim who whispered that Long Shanks was walking around the docks.

As we passed the sheds we saw the aftermath of what had happened. There was debris strewn everywhere and the timbers that used to hold the sheds up were burnt and as black as charcoal. Truckers stick together and when they decide to do something they certainly don't mess about.

We turned right and headed towards the quay where the cranes were erected. I spotted Long Shanks in the distance.

'Who are you? What's your business here?' he demanded when we approached him.

'Our business is you!' Benny replied quietly. 'We're here on behalf of all the drivers whose lives you have ruined by not paying them their wages, not to mention what happened in number eight shed. We're here to clear up a few matters.'

Long Shanks looked straight into Benny's eyes. 'Number eight shed! I don't know what you're talking about! Anyway, I don't have to stand here listening to this from a couple of upstarts.' He looked as though he was shaping up for a scrap.

Without more ado Benny got his retaliation in first, lashing out straight at his stomach. Long Shanks went ballistic, but he was outnumbered and came off worst.

'Come on Ben. Let's get out of here,' I said when we had vented our pent-up anger.

Back at the security office I thanked Jim for the tip-off. 'I've not seen no one and I don't know nuffink,' Jim said as we left.

As we walked over to Long Shanks's car, Benny grabbed the windscreen wipers and bent them clean in half. 'Did you like that?' he quipped.

'You do make me laugh, Ben. Long Shanks can buy another set for a couple of quid.' Ben's response was to kick the headlamps clean off the car. 'That's it. He can buy them as well,' he said.

As I drove up the lane I started to feel a little guilty. 'Ben? Have we gone too far?'

'That's the bloody trouble with you, Rich. You're too chicken-hearted. Just think of the drivers in Hull, Manchester and Kent who are suffering through that toss-pot. What we have just done is nothing compared with that.'

'You're absolutely right,' I answered.

As we drove along the Gravesend road we stopped off at the Sir John Falstaff pub near Charles Dickens's house on Gad's Hill where we treated ourselves to a couple of jugs.

As I drove through the Medway towns the sun appeared to be shining extra brightly and there was a lovely aura around us. 'Ben, I've got this feeling everything is going to turn out OK,' I said.

Ben patted my knee. 'It's funny but I can feel it too.'

At the farm we were greeted by Jenny's smile: 'Hi Rich, Hi Haggis.' It had been years since she had called us uncle but 'Haggis' had stuck. It always made me smile.

Len came by dangling a freshly killed chicken for the house. 'Could I have a word with you both?' he said. 'Let's go up to the house. We can talk better there.'

As we walked through the back door, Jean looked up. 'I'll go and make us all a brew. I expect you could both do with one.' Len hung the chicken on the back door, and then joined us.

'Jean's told me how you intend to keep going. But there's a recession on and you're going to need some capital, aren't you. How much do you think you'll need?'

'I hadn't really thought of that,' I replied.

Len looked over at Jean. 'How much money does North Kent need to get kicked-started?'

'I would say just off the top of my head eight thousand pounds. Even if we get the work we'll have to wait until the money comes in. But they're good workers, Len. We're backing them even though we have very little money ourselves.'

Len stood up from the table. 'OK, get working. I'll have the cash ready in two days time. All I want back is the amount I lend you. OK?'

'You can't do that, Len,' I exclaimed.

'Chuckles and I started this farm up together and we managed to get if off the ground. The truck he bought made more money for the both of us, so I don't want any more said about the situation.'

Chuckles, fully recovered, left for his new job later that week. It was a temporary position and the future was uncertain but he had friends to stay with and the money was good. If all went well he intended to move his family to join him later. We were sad to see him go, we were going to miss his irrepressible sense of humour.

This period of the 1980s was the worst for the transport industry. What had happened to us left a bitter taste and gave us even less respect for rules and regulations. We had to make the job pay. If we got caught it was down to us.

Within a few days Jean was managing to find us work and we were back in the saddle, punching our way to all different parts of the country. One day when I returned from a trip to Woking Jean told me that an unusual job had just come in.

'Drop your trailer, Rich. You're to go over to All Hallows and hook up to a caravan trailer. It's forty-two foot and already loaded for Windermere.'

'The rate?'

'Better than I expected,' replied Jean. 'It's excellent.'

After I unhooked my trailer I went looking for Len. 'Can I borrow your HGV licence please?'

'Certainly,' he answered with no questions asked.

Soon I was driving towards the north Kent coast and in just under the hour I had pulled in at All Hallows on Sea. The campsite was beautifully laid out – a far cry from how I remembered it as a young boy in wartime. Then there were guns on concrete stands and the place was covered in barbed wire to repel an invasion.

As I got out of the cab I was approached by a 25-year-old Jack the lad who had a wad of pound notes in his back pocket.

'My truck's blown its engine,' he said. 'I've been working it so hard. But we've got to get this caravan up to the Lake District as soon as possible.'

'I only hope my rig fits your trailer,' I replied.

'Have faith. It will.'

When I checked the couplings my luck was in. They were the bayonet type and I was able to hook up. It wasn't a bad little trailer but very light. It probably weighed only about three and a half tonnes, designed purely for caravans.

'Driver, I want you to get this up there as soon as you can,' the lad said. 'I've had a lot of problems at the other end and my truck breaking down hasn't helped the situation. What's the earliest you can get there tomorrow?'

'It's about half-three now,' I said. 'You tell me what time you want me there.'

'Seven in the morning,' he replied, with a laugh.

'That's no problem for me.'

The lad handed me a fiver. I thanked him and pulled off the caravan park. As I left the gate, I took my old tacho out and put a new one in with Len's name on it.

There was so little weight in the load that it felt like driving an empty truck. However, the caravan was just under nine feet wide which meant that I couldn't see through my mirrors very well.

When I cruised up the notorious drag at Swanscombe cutting I was able to keep my speed to around sixty-five. My rig didn't even know the hill was there; it was effortless. I had to stop just before the entrance to the Dartford tunnel and wait for an escort. They soon arrived and in no time at all I was driving through with the Land Rover behind me flashing its yellow light. At the end of the tunnel it pulled off and I carried on punching up the M25 on a glorious afternoon with bright sun in a blue sky.

I made up my mind that my next stop would be the Blue Boar on the M1. I kept my eyes open for the wooden tops as I made the Volvo tramp along. Then suddenly I had the most terrifying experience of my career. When I came into the open space on the motorway the trailer which held the caravan started to lean over as the wind caught the light load. 'Bloody hell!' I thought as I gripped the steering wheel. It unnerved me so much that my arse started to bite the buttons off the seat.

I started to brake gently until I came to the part of the motorway which had trees either side where I gave the truck more gas. I learned quickly how to pull a trailer with a static caravan on the back. Going past Luton there were a lot of open spaces where the caravan and trailer kept leaning over and rocking vigorously. By now I was calling it all the names under the sun.

At the Blue Boar one of the drivers was talking about the good old days. 'Driving's finished now,' he said.

'But at least we don't have to rope and sheet in all weathers like the poor sods used to,' I replied.

Once more back in the saddle I drove north. With time getting on there was hardly another truck on the M62. Of course, I was drowsy, but I began to recuperate once I left the boredom of the motorway.

I cruised up the A590 on the outskirts of Kendal and in no time I was on the A591 pulling in at the Buxton caravan site. It had been a long haul to Windermere and that was after driving to Woking and back. It didn't take me long to remove the tachograph, draw the curtains, get undressed and climb up into my bunk. It was a quarter to twelve and as soon my head hit the pillow I fell asleep.

I slept like a log until seven-thirty when I heard voices outside. Within a quarter of an hour I'd climbed down from my rig. One of the workers brought an old tractor round, and a couple of them dropped the ramps to keep the caravan balanced while they pulled it off the back of the trailer.

I took full advantage of their facilities by having a shave and a lovely hot shower. Even if I was driving like a cowboy I didn't want to look like a hobo. I always wanted to give a good impression – especially if I was stopped by the ministry.

I said goodbye to the lads, jumped up inside the cab and wrote the tacho out, using my name this time. If I had been stopped by the law I would have told them that I had just changed over, as there were two drivers to one truck. I suddenly thought that I'd better check the unit and trailer. The oil and water were all right and the trailer appeared OK although it was made mostly of aluminium and looked extremely flimsy.

I was about to climb back inside the cab when a car pulled up driven by a dapper old gentleman with a goatee beard, piercing blue eyes and a ruddy complexion. 'Are you going south?'

'Yes,' I answered. 'Why do you ask?'

'I've got a ship's propeller that needs centralising and it has to be in Birmingham as soon as possible. My workers didn't think they'd get if off last night. When I saw you parked up I thought it would be quicker for you to deliver it as you're already in the vicinity. It's at Whitehaven. We have a large crane to handle the propeller and it's to be delivered at Cox Shipping Engineering, down in the Black Country.'

'What are the rates?' I asked

'Eighty-five pounds.'

The old boy didn't dither as I followed him into Whitehaven. When I backed my rig alongside an old quay, he soon got his men organised and in minutes they were lowering this rather bulky propeller onto the back of my trailer. Being a low loader it looked lower than low. The only material I had for securing it was rope though really, as the load was all steel, it should have been chained. However, because of its dead weight I figured it wouldn't go anywhere.

I was very pleased when the old boy paid me up front, after which I made my way to Birmingham. I hadn't travelled far when the rig felt like it was running on kangaroo juice. One second the lorry was pulling and the next it was braking.

I kept her going until I found a space wide enough to pull over. I had a good look around and then noticed that the trailer was slightly bent in the middle because of the weight of the load. Every time I ran over a bump it put extra strain on the chassis and stretched the hand-brake cable – and put the trailer brakes on.

The trailer was so low that I couldn't get underneath it. The handbrake on the side of the chassis wasn't like the modern ratchet but more like a car which pulled a rod brake. I pulled the split pin away from the bolt and undid the nut, pulling the bolt away. This released the rod from the handbrake. It wasn't very robust but this trailer was only meant to carry caravans and I hadn't a clue how much the propeller weighed. The power of the 290 Volvo engine made it impossible to tell what weight I was carrying.

After giving the trailer another going over I thought it did look a bit sick and it certainly didn't like the propeller. However, I decided to have another go.

As I travelled along the A66 I wound down the window and listened. She was groaning, creaking and making one hell of a noise. By now I was getting concerned.

'Rich,' I said to myself, 'I think you've dropped a bollock.' There wasn't a thing I could do apart from turning around so I decided to carry on to Birmingham. After all it was all motorway once I was back on the M6.

I pulled in at the Lancaster services where I felt the hubs on the trailer. I was concerned to find they were really hot. The trailer chassis looked more twisted and even a bit distorted.

'I just hope the grease in the bearings doesn't melt,' I thought. 'The bearings will seize up and chew the ends of the axles. A very expensive job.' All I could do was keep my eye on the wheel bearings and watch just in case they started smoking.

As I sat and ate my meal I felt ill at ease about the trailer. 'Why do I put myself in these situations?' I asked myself. 'I must be a fool. I've got enough problems on my plate without adding any more.'

As I tramped on down the M6 I kept my eyes peeled in both mirrors, watching out for smoke from the hubs. Lady luck was certainly with me that day.

When the crane driver at the engineering works lifted the propeller shaft off the trailer it gave out a groan as if to say, 'thank goodness I've got rid of that.'

'Do you know roughly how much the propeller weighs?' I asked the crane driver when he climbed down from his cab.

'About nine and half tonnes'

'Can you tell just by looking at it?'

'There's a meter in my cab,' he laughed.

Nine and a half tonnes on a caravan trailer! No wonder the noise was ear-splitting. 'You've got to laugh,' I said. 'It's the only thing that keeps you sane.'

I drove into Corley Services, connected the cable rod back onto the handbrake and inspected the trailer just to make sure everything was in order. Now it was unloaded it didn't look too bad.

I didn't stop any more until I arrived at the Esso garage on the A2 in Gravesend. I jenked up and had a chat with Pamela who was behind the till. She was a pleasant young woman who all the drivers knew well. They always teased her and told her jokes. She'd put her hands over her ears pretending not to listen but she'd be laughing at the same time.

When I rang Jean she said that there was a stack of work and that my trailer was already loaded. I put the phone down, climbed up into the cab, took the trailer back to All Hallows on Sea and headed back to the farm.

# Chapter 11

## Driving Night and Day

WE made our money by doing work that the larger transport companies didn't want. We couldn't afford to turn it down.

One nice little earner at Barking was delivering 45-gallon drums of lube oil to back-street garages all over the Home Counties. Between the three of us we must have covered nearly every one. Manoeuvring our 32-foot trailers around the tight back turnings was often difficult and we counted ourselves lucky when we had as many as three drums for one delivery. We were supposed to take the empties away but usually we didn't bother. It was all too time-consuming.

We found that the quickest way to unload was to throw an old truck tyre on the ground, roll the drum off the truck onto the tyre and then toss the tyre back onto the truck. Get the ticket signed, job done.

Leaving Barking one day I decided to drive to Brighton first as I had ten drops in that area. I drove all around the Kent coast making the drops as I went, eventually arriving at Brett's in Whitstable, an aggregate company that had a massive fleet of eight-wheelers. I delivered my last two drums there.

As I sat in the cab I counted forty-two delivery notes.

Although the rates were good we were all brassed off with the work and driving. We lost the spot-hire work after a week because we didn't collect the empties but none of us were bothered about that. I often wondered what Chuckles would have said in our position.

The next job Jean found us was one that brought in a lot of much-needed income. We started by reducing our trailers to flat-backs, then going to a toothpaste company in Manchester. There we loaded enormous plastic tanks which were on a chassis with big iron lugs on the side so we could rope them down onto our trucks. Our job was to go to Immingham and load these tanks with 'petrolab' for delivery to the Manchester factory's storage tank. This was always to be kept more than half-full because the chemical was vital to the operation of their factory. If necessary we were to put on extra drivers and operate day and night.

The three of us soon got ourselves motivated. All the documentation and rates of pay were sent direct to Mrs Sorter, secretary of North Kent Transport who arranged for money to go straight into our bank accounts. We had no administrative worries. Any problems were dealt with by Jean – we were just drivers doing a day's work for North Kent. Jean dealt with the problems while we earned as much as possible and saved hard towards our new lives in New Zealand.

From the toothpaste factory the three of us made our way to Eccles and soon we were on the M62. Little did we realise then that we'd be on this route so much that we'd know every bump and groove in the road.

It was a fair old trip to Immingham dock but once we were there it was a swift load. We just had to make sure that the outlet valve was closed, climb up onto the gantry, insert the loading arm into the pot, put the delivery notes into the machine and press the button. When it reached the amount that was on the ticket the electric pump would stop. Then we retrieved the pot, closed the lid on the tank and took the ticket from the machine. Back on the road once more.

When we arrived in Manchester we would report to the laboratory where they would give us a bottle, then we would proceed to take a sample of the liquid. The man in the laboratory didn't take long to test it and would tell us to pump the load off. We would pick up the customer's hose and connect it to the tank on the truck, and then the driver would climb the ladder to the top of the road tanker and open the lid. If the driver didn't do this the pump would suck the sides of the tank in and that would be a major disaster. Then he would climb back down the ladder, unwind the valve, press a button and the customer's pump would automatically start up and draw the liquid from the tank. When the hose started to shake, the driver knew that he was empty so he would close the valve, turn off the pump, unhook the hose, close the lid and drive away.

After two or three days we realised we could not waste time. We were like bumble-bees just driving backwards and forwards doing the work to the best of our abilities. Work meant money. However, going from Manchester to Immingham and back was a hard day's work so we decided to find a base with good facilities where we could get a decent meal. We chose Hill Top at Blyth, Notts. This was about eighteen miles off route but it was well worth the extra because it was a day-and-night café. So whatever the time we were always sure of a shower and a decent meal.

We knew that we would not always be running together because of the delays of loading and unloading, so the only time we could guarantee that we would meet up was when we parked up at night. At one time the factory was on full shift which meant that the three of us had to work day and night. Sometimes it was so bad that we didn't even bother to put a tacho in. We were doing the work of six drivers every twenty-four hours. This carried on for at least a couple of weeks after which it eased back to one load each per day, which was acceptable.

Jean's spirits began to rise again. 'All the payments are on time,' she said, 'and you know the rates are really good. We're making money at last.'

'Wing Nut is just about exhausted.'

'That's one problem we do have, Rich. His wife is getting disheartened, she hardly ever sees him now he's away working all the time.'

'OK. We'll work something out.'

At our evening meal we agreed that Wing Nut deserved a few days off. Benny and I decided to cover this by adding an extra half-load a day to our own workload. After a few days of this the work not only became a nightmare but we were sitting behind the wheel much longer than we should. If we had been stopped by the wooden tops we would have been booked straight away and we couldn't have done a thing about it. We were making the same mistake we made running ragged for Shorne Mead.

Even when Wing Nut returned we were still over-stretched. On one occasion when I was following him on an almost deserted M62 we came to some roadworks. Before I could do anything he wandered into the cones, scattering them everywhere, blocking one side and even spraying some over the crash barrier onto the other side of the motorway. The event certainly woke Wing Nut and me up and we carried on full-steam ahead.

When we pulled in at Manchester to pump off we checked his rig and trailer to see the damage. Cones had jammed themselves under Wing Nut's front axle and one was even lodged under his mudguards. He said that after a few miles they had worn down to half the size before eventually flying off. One of the mudguards on the trailer was distorted badly. I wouldn't have imagined that plastic could have done all that damage. Thankfully the air lines were still intact.

'Thank goodness no one was hurt, Rich.'

'And no one saw us!'

As Wing Nut coupled up to unload I started the customer's pump and could hear the liquid running through the hose. 'We can't carry on like this, Wing Nut. Someone's going to be injured. Today we'll just go back to Immingham, load and park up at Blyth. Let's just do the one load.'

I then went to see the laboratory man. 'Do me a favour, Dave,' I said. 'Tell Benny that when he's tipped here he should go straight back to Hill Top, and not bother to load at Immingham.'

On my way back to Blyth I saw poor old Benny looking rather glum at the back of the queue held up by the maintenance men picking up the scattered cones on the other carriageway.

When we all met up at Blyth we discussed how to run our trucks better and more safely.

'Is Side Valve available for work?' I asked Jean on the phone. 'We've decided to change over at Ferrybridge, between Manchester and Immingham, and run a trunk service. Each driver will have a week delivering in Manchester, then the following week change over and have a week in Immingham. If Side Valve comes in with us we'll even be able to deliver six loads a day if we need to.'

'I'm sure he won't turn the opportunity down, even if you get him working flat out seven days a week. I'm glad you've phoned,' Jean carried on. 'I've some more good news for you. We've already paid Len back and all the bills are up to date. Any work you do now is profit.'

I gave Jean my love and then the three of us at Hill Top got down to sorting out how we actually got our new system working.

'I think we should start early,' I said, 'say three-thirty in the morning for all of us from here. If we do that we should be parked up by mid-afternoon.

The first week Benny and I will do the Manchester run. Wing Nut and Side Valve will bring loaded trailers here and we'll swap overnight so in the morning Ben and I drive straight to Manchester to unload. Then we do two more trips from Manchester to Ferrybridge and back, swapping our unloaded trailers for loaded ones at Ferrybridge each time. Finally we come back here empty in the afternoon.'

'The drivers on the Manchester run are going to be doing extra mileage,' remarked Ben.'

'Yes, Ben, but the other two will be using less fuel. And we'll swap each week.'

'We don't know how long this job will last,' Wing Nut added. 'Let's make hay while the sun shines.'

'What are you going to do with all your money, Wing Nut?' asked Ben.

'You're self-employed on the books,' I added. 'That means you'll have to declare what you've earned to the taxman.'

'You don't think I'm going to worry about such trivialities do you? The taxman didn't worry about me when I lost all my hard-earned cash to Long Shanks. I could have been sleeping on a park bench for all they cared.'

'Just be careful,' I said. 'The revenue always catches up with people in the end.'

'Oh, I'll be careful, all right,' Wing Nut answered with a wink. 'Well, how about some tea,' he continued, looking straight at Benny. 'Let's see the moths flying out of your pocket, you tight-fisted old goat.'

Benny must have regarded Wing Nut as a good friend or else there would have been fireworks. We teased Benny a lot, but it didn't do to get on the wrong side of him. Perhaps it was the news that we'd soon have enough money to go to New Zealand that put him in such a good mood.

When I rang Jean back later she told me that Side Valve was with her. She handed him the phone.

'Would you like to do some work with us?'

'Too right I would. There's no money in the job down this part of the world.'

'Nine tomorrow morning at Hill Top. OK? Remove the curtains and all timbers so it's a flat-back trailer.'

'Will do. I can't wait to be with the lads again.'

The system worked fine for the six loads with four trucks although we were having to average three hundred and forty miles a day. By late afternoon we were having a shave and a shower before tucking into a meat pudding, potatoes and carrots with lashings of gravy followed by apple pie and custard. It went down a treat.

Within a week we were able to cut a load out because the customer was unable to take it. We were on top of the job. The four of us had never had it so good. We had good trucks, plenty of sleep and at the end of the day we had one goal in mind – to earn money.

As usual Jean handled all the paperwork. She was so efficient that I don't know what we would have done without her. Every Friday I would go to a bank near the toothpaste factory, where I would draw out two hundred pounds to split between the four of us. That saw us through the week.

I had one near miss during this period. It was four-thirty on a cold morning when I was leading Benny on the M62. I was just about to turn the radio on when suddenly a huge object was bouncing towards me so quickly I couldn't make out what it was. I swerved over to the right, and with more luck than judgement it whizzed past me.

The lorry in front of me pulled over onto the hard shoulder and Benny and I pulled up behind him. He had a DAF unit on the back of his lorry so I presumed it had broken down at some time. 'Bloody hell,' said the driver. 'I'm in the shit now.'

The DAF unit had a windbreaker on top of the cab. Whoever had loaded the truck had pulled it up backwards on the low loader so that the aperture on the windbreaker faced the front rather than the back. It acted like a huge sail. The whole lot had been completely ripped from the cab, leaving four large holes in the roof. Fibreglass was strewn all over the road.

The driver looked as sick as a cowboy's horse. Unfortunately we couldn't do anything to help and we had to get moving as we didn't want to delay our change-over.

When we all met at Ferrybridge later that morning the other two wound the wheels down on our trailers. We disconnected the air hoses, pulled the pin then pulled the bar back and unhooked the fifth wheel coupling. Benny and I pulled away from the dropped trailers and the other two reversed onto them, then we did the same on the empty ones. We really worked fast so that the complete change-over took six minutes exactly which wasn't bad.

We spent twenty minutes in the café, then Benny and I drove back to Manchester, while the other two returned to Immingham.

We knew this good work wouldn't last for ever and weren't too surprised when Dave from the laboratory told us that they were having new tanks built. 'These won't be fitted with their own pumps so the chemical will require specialised transport. It will be more efficient and economical for us.'

'When's this happening?'

'Over the next two or three months.'

While we were uncoupling the hoses from the tanks, Benny looked at me and winked. 'That'll work out just right for us. We'll keep running the trucks on a shoe-string and we can earn quite a bit of money in that time.'

But when we got back to Hill Top we found that Side Valve was less impressed. 'I thought this job was too good to be true,' he exclaimed. 'I'm earning good money now and the job's easy.'

'Easy!' said Wing Nut. 'Driving every day flat out. No wonder we're whacked most of the time. You must be cab-happy.'

We carried on working seven days a week and hardly thought about Dave's information until Side Valve and Wing Nut said that they had seen the new storage tanks being installed.

Another month passed with no news until the inevitable happened. 'Unfortunately, tomorrow is the last day for you and the lads delivering loads to Manchester,' Jean said on the phone. 'So I'll see you all on Saturday morning.'

I broke the news to the others in the Hill Top café. 'I've had a thought,' said Side Valve. 'We can do six loads into Manchester with four trucks, so why don't we do eight loads with the four trucks?'

'What are you talking about?' Benny exclaimed. 'That's impossible. The tanks at the other end won't hold that amount of liquid.'

'Oh, we don't have to load our tanks to the brim. They've never had any problems with us, and we've given them full service. One day on the fiddle won't hurt anybody.'

'I'm not sure I like this,' I said.

'Here we go again,' said Benny. 'This is a good idea. Let's just do it.'

We turned in early that night because delivering eight loads meant driving a lot of miles and we didn't want to fall asleep at the wheel.

The first two loads we delivered were full and then we started filling our tanks to the halfway mark, delivering about ten tonnes a load instead of twenty. All the time Dave was signing the tickets we carried on regardless.

At the end of a long and arduous day Benny and I tipped the last load. We went to shake hands with Dave. 'You've given good service,' he said. 'It hasn't gone unnoticed by the company. It's a pity you couldn't afford your own tankers. You could have had the contract.'

When we ate that night Side Valve was like a cat with nine tails, rubbing his hands with glee. 'Eight loads, eight loads,' he kept repeating.

'Well, we've certainly got some front,' I thought to myself. 'Maybe it's the bad experiences we've had that have caused it.'

The following day we had a leisurely breakfast before making our way back to Kent. The A1 was clear, the sun was shining and there was plenty to see from the cab. It wasn't long before we were pulling in at the farm.

As we drove past Jean's house we blew our horns, then we drove down to Chuckles's shed. When we climbed down from our rigs, we were immediately greeted by the smiling faces of Jean and Jenny. They certainly gave us a lovely welcome home.

176

'This is a nice surprise,' said Jean. 'I wasn't expecting to see all four of you looking so well – and clean-shaven too. I was expecting you to be on your knees after working flat out for fourteen weeks. Come on in and I'll put the kettle on for a cuppa.'

'It's about time. We've been here at least ten minutes,' I teased her.

'I see you haven't changed,' she laughed. 'Cheeky as ever.'

We sat down at the familiar table where there was still an empty chair for Chuckles. 'I've got you four loads out of Robertsbridge on Monday through Jenson's Transport of Rye,' Jean said over her shoulder while she was pouring the tea.

We looked at each other dumbfounded. 'Bloody hell, Jean!' I exclaimed. 'We've been working non-stop for the past three months.'

'I could always cancel it.'

'You can cancel their work but not mine,' Side Valve piped up. 'I need the money.'

'I've been taking care of your wife,' said Jean. 'She's had at least a hundred pounds a week to live on – and it hasn't come out of your earnings.'

'How do you mean?'

'Well, the toothpaste company has been paying us twelve pounds a night out for every driver. I told them that there were six drivers operating, so we gained twenty-four pounds a night, seven days a week.'

'I see Chuckles taught you well,' I said as the others nodded approvingly. 'Don't worry, Jean. We'll do the Robertsbridge job.'

When we had drunk our tea I was keen to get to the

home we hadn't seen for three months. However, Benny had other ideas. 'I won't be leaving here for at least another two hours because I want to wash the truck down.'

I looked at him disdainfully. 'You're right,' I said. 'I had better wash mine down as well.'

Side Valve also decided to stay but Jean ordered a taxi for Wing Nut as he wanted to get home. I took his keys from him and told him that we would sort his truck out.

I hadn't realised how hard it was to reassemble the taut liners especially having to do it on all four trailers. By the time we had washed them all down it was six o'clock in the evening.

Benny and I went for a Chinese then made our way home where we had a terrible job trying to open the front door which was blocked with mail, most of it junk.

On Sunday we were woken by someone banging on the door. When I peered through the curtains Wing Nut and Side Valve were standing on the doorstep. 'We thought you were never going to open the door,' said Side Valve. 'It's freezing out here. Have you only just got up?'

'Why? What's the time?'

'It's the afternoon – two o'clock.'

'Blimey. We must have been tired.'

The four of us walked in a flurry of snow down to the curry house. There we had our usual bantering fun, mostly at Benny's expense.

When we arrived back at the house Benny and I went through the mail which was still piled up behind the door. The most important letter was one from New

Zealand House in London, inviting us for interview in eight days time.

'You know what,' said Benny. 'It's typical the way things work out. This country is gradually coming out of a recession and now we're leaving it and going to set up home in another.'

'Well, we could always change our minds,' I responded.

'Not bloody likely, laddie.'

'Don't start getting yourself in a tizzy, I'm looking forward to it as much as you.' That was true. I was beginning to feel really excited.

When we arrived at the farm on the very dark Monday morning the weather was bitter. As we pressed our starters all four rigs flew into life. I revved mine for a while to build the air up.

In no time at all we were driving through Maidstone and following the A229 all the way to the A21. We eventually pulled in at British Gypsum at Mountfield, just outside Robertsbridge. Although the loads were one hundredweight bags of plaster it didn't take the gangs long to load us for Weston-super-Mare. We just slid the curtains back and got our delivery notes.

We stopped at the first services on the M4, aiming to be in Weston-super-Mare at about three p.m.

No sooner had we climbed back into the saddle than Side Valve was pulling out. The race was on. We really pushed those rigs all the way down to Bristol. Side Valve led us straight through the centre of the city, along the way turning onto the A38 that took us into Weston-super-Mare.

Eventually we found a huge complex in Milton. We

could see by the amount of plasterboard that was stacked in the building that we were at the right place. We were approached by the foreman who told us to line up one behind the other. The men climbed up onto the backs of the trucks and handed the bags down to their workmates who stacked them alongside the wagons. Naturally we pitched in and helped. When we had unloaded the four trucks we were covered from head to foot in white powder.

I retrieved my brush from the cab and brushed myself off. Side Valve brushed my back and we took it in turns to help each other. Side Valve was paying close attention to Benny's bottom for a laugh. Benny, being Benny, didn't see the funny side of it at all. 'I shall bury you with that brush stuffed up your . . .' I heard Benny shout.

'Oh, for goodness sake, Ben, lighten up,' shouted Wing Nut. 'When will you learn to take a joke?'

As we travelled back through Bristol, my brain started to go into overdrive. By the time we arrived at the Leigh Delamere services on the M4 I had made up my mind. We were sitting down to a substantial dinner when I broke the news to the others. 'I'm calling it a day,' I said. 'This is my last trip.'

'That's a bit sudden isn't it? It could be ages before we go to New Zealand!' Benny exclaimed.

'I can always get a job driving a van for a short while. I look at it this way. We were right to run the trucks on a shoestring but now we've got to get them sold, otherwise we've got to pay out for tax, insurance, MOTs and servicing. I'm not going to waste the money that I've earned.

'We've been really lucky getting away with running bent but our luck's going to run out sooner or later. If we get fined it will be all our hard-earned money down the drain. In any case, we deserve a break.'

The discussion went on until we had finished our meal. Wing Nut was still anxious. 'There's nothing to worry about,' I told him. 'You can have the choice of Benny's or my truck, and if Jean carries on she may want you to drive Chuckles's rig. Either way you'll be all right.'

'Why don't you work for me?' Side Valve suggested. 'I'll buy one of the rigs for you to drive.'

Later I phoned Jean. 'I don't blame you,' she said. 'I've had enough too. I'm getting a bit concerned about the taxman and the ministry. Probably best to get out while all is well. No work's come in so far today, and I won't take any more. From now on, North Kent doesn't exist. Have a couple of days off and I'll see you down on the farm Thursday morning. I'll square up with all of you then.'

So that was that. We made our way back to the farm and left our trucks there.

# Chapter 12

# New Zealand

IT was great having a couple of days off. I even got some shopping done so that the cupboards were stocked up with food again.

On Thursday Ben and I sat in Jean's scullery drinking tea and reminiscing about the laughs we had had as North Kent. 'It looks like we may be moving to Birmingham,' said Jean, 'I don't want to but the job is going well for Chuckles and it would be nice to have the family together again.'

When Side Valve and Wing Nut arrived we moved to the office so that Jean could sort the money out. 'Now, Wing Nut,' she said, reaching for a box full of ten-pound notes, 'after deducting the money I've been paying your wife you're due four hundred and fifty pounds a week, and there's a bit extra for this last job in Weston-super-Mare.'

'You couldn't hope for better than that, could you?' Wing Nut responded.

Jean then gave Side Valve a large pile of neatly typed papers itemising all the work he had done. 'You should find everything in order,' she said. 'The diesel's all paid – the receipts are on top – and I've taken out our ten per cent management fee. That still leaves you nine thousand four hundred pounds.'

Side Valve looked really pleased. 'I've a favour to ask you now,' he said. 'I'd like to buy one of the trucks.'

'You can have all three if you like,' I replied.

'How much, Rich?'

'Seven thousand.'

'Make it six and you've got a deal.'

I glanced over at Jean and Ben who nodded in agreement. 'Done,' I said. 'We'll have to remove the name of 'North Kent', though.'

Then we turned to some of the other jobs to be done in winding up the company. 'You'd best get all the tachos and other paperwork together and burn the lot. Then there'll be no incriminating evidence,' I said.

'That's no problem,' Jean replied.

'Oh, I see,' Wing Nut piped up. 'I've been working for a governor who's a crook, have I?' We all cracked up with laughter.

Jean was the first to recover. 'You realise you won't be able to leave the trucks here any more, don't you.'

'That's OK. I'll take them away on Saturday. We'll shoot off now and make room in my yard,' replied Side Valve.

'See you Saturday morning,' called out Benny, 'with the cheque for the trucks.'

'Trust a Scot to remember!' Side Valve shouted as he closed the door.

Jean then turned to Benny and me. 'Here are your cheques,' she said. 'You'll find they come to just under sixteen thousand pounds each.'

'I can't believe it,' said Benny. 'Four months ago we were penniless.'

'We'll need all the money we can lay our hands on when we set up in New Zealand,' I replied. 'We'll bank the cheques now in case the immigration authorities want to verify our incomes. Then is there anything we could do for you, Jean?'

'Yes. If you wouldn't mind, Chuckles has said I can get rid of his shed, he doesn't need it any more. Could you knock it down for me please?'

When we came back from the bank we opened the shed door and stood amazed at the amount of junk Chuckles had stowed away. There were all makes of wheels, including tractor ones; all Chuckles's tools, even welding gear and an anvil. There was every conceivable lorry sheet with different companies' logos. It was just like Aladdin's Cave.

'Where on earth are we going to put it all, Rich?'

'Well, Side Valve bought the rigs and that includes everything that goes with them. We'll throw it all on the back of your trailer for him to collect.'

Benny drove as near to the shed as was possible. We threw the whole lot on the trailer, loading it pretty well up to the hilt.

After that Benny tried to dismantle the shed the easy way. 'See me reverse into it, Rich, and I'll push it over.' However, when he tried it didn't budge an inch.

'I'm going to take a run at it,' he shouted. 'It should move then.' He drove the truck up about ten feet, dropped her into reverse then gave it full throttle. I ran for my life – talk about full-steam ahead. The man was mad. The whole lot – trailer and all – went straight through the shed and out the other side, leaving dust and rubble everywhere.

When he eventually pulled forward, I could see that the number plate was completely bent and the rear lights were smashed.

Jean came rushing out of the house thinking that a bomb had gone off. But when she saw what had happened she just collapsed with laughter, saying it looked like something you see in the movies. She walked back to the house still laughing at the sight.

Benny was slower to see the funny side of it. 'I didn't mean to damage the truck. I like Side Valve. I'll put it right.'

We threw all the old timber on the back of my truck and drove it down to the bottom of the farm where we could burn it safely. When we had finished clearing up we called Jean out again.

'That does look better,' she said. 'The shed was a real eyesore. It must have been there for more than forty years.'

The following day Benny and I came back to the farm to remove the name of North Kent from the top of the cabs. We straightened out the number plate on Benny's truck, then put a rope around the crash bar at the back of the trailer, tied the other end onto the back of my vehicle and pulled the crash bar out. 'That'll have to do,' said Benny. 'We won't worry about the lenses. He's getting a good deal, anyway.'

I didn't enjoy the final meeting on Saturday. It was a heart-rending day during which I kept thinking that if Benny wasn't so keen on going to New Zealand we could have ridden out the storm and made a fresh start here.

Side Valve handed us two thousand pounds each in cash for the trucks and that was the end of North Kent.

Our interviews at New Zealand House went through smoothly, but waiting for our papers were the longest three weeks that we had ever known. Every time the letterbox rattled we were at the door before you could say Jack Robinson. Eventually our confirmation letters arrived and we had a date.

Then we bade our farewells to Jean, her children, Len, Wing Nut and Side Valve. 'We'll ring you, Jean, and let you know we've arrived safely,' I said. 'We'll write when we get settled.'

'I'll believe that when I actually get a letter,' Jean laughed.

We had to say our goodbyes to Chuckles over the phone which wasn't exactly what I would have liked.

We paid the landlord what was owing to him, then took a trip to see my Mum and Dad who wished us the very best of luck in our new venture. 'Just think,' said Dad, 'if we want free board and lodging for a holiday in New Zealand we'll have somewhere to go.'

'It works the other way round too,' replied Benny. 'We'll expect free board and lodging when we visit you.' That made them chuckle, though poor Mum was tearful and Dad was choked as they waved us off.

We sold the car for some cash, checked our documents were in order and started our last night in our house. I felt apprehensive about going into the unknown. 'I can't believe it's happening,' I said. 'The months have just flown by.'

Before Benny could reply there was a bang on the door. Side Valve and Wing Nut were standing there with a crate of beer. 'We can't let you go off without sharing a couple of jugs – and if you like we'll take you to the airport tomorrow.'

We relaxed as we sat downing a few pints, talking about the past and our futures in New Zealand. Then Side Valve got serious and looked me straight in the eyes. 'The real reason we are here—'

'Wait for it. Here it comes—'

'Is that you filled one of those trailers up with a load of bloody rubbish. Now I've got to get rid of it.'

'Not me,' I said. 'It was him,' pointing to Benny.

'You decided to buy North Kent out lock, stock and barrel. Well, that included all the company's equipment on the back of the trailer,' he said, dead-pan.

'You know full well, Walker, that I've no room for that lot in my yard!'

'I'll tell you what I'll do, laddie. I'll borrow one of your rigs and take the lot down the tip for – let me see – five hundred quid. Cash in the hand, mind.'

'You get knotted,' Side Valve replied.

'Now, who can't take a joke, then,' Benny said with a smile.

The flight to Auckland was a long one, leaving us so tired that it took a full two days to recuperate. Then we bought a Toyota saloon and started to drive round North Island.

After three days of marvellous scenery we were nearing the logging country around Taupo. When I looked across lake Taupo – the largest lake in New

Zealand, Benny informed me – I could see snow-capped mountains. I fell in love with this magical place.

'I don't need to go any farther.' I said. 'I really want to settle here.'

On the way back to Auckland we saw some beautiful rigs. I slowed down so that we could get a better view of them as they overtook us. They were 'wagon and drags' loaded up with pine trees. The trailers were eight-wheelers, about thirty feet long and they were being pulled by powerful-looking American eight-wheeled trucks. They made a crackling sound through the exhaust pipes as they punched south. Ben and I looked at one another and nodded approvingly.

After revisiting Auckland we made a trip down to South Island. I thought I knew Benny well but I was learning more about him all the time. He was an absolute mine of information about New Zealand. On the ferry across Cook Straight he pointed to a place on the right. 'That's called Glasgow Bay,' he said.

'How do you know all this?'

'I had a good teacher in Scotland. We're not like you riff-raff down in the south.'

We drove through endlessly breathtaking scenery until right in the south we stopped at Balclutha, a decision that turned out to be the best thing that we could have done. It was there that we met Maria and Louise, two charming Kiwis who were on holiday from North Island. The following day the four of us travelled on to Catlins.

None of us wanted this holiday to end. By the time we had reached the ferry back to North Island I had become fond of Maria and, although Benny hadn't said anything to me, I was sure he was smitten with Louise. We dropped the girls off at Russell.

'This has been great,' I said to Benny that night, 'but the money's slipping through our fingers like desert sand. We ought to keep to Plan A which means buying a house soon.'

It took us four more weeks of bed and breakfasting to find what we were looking for – two properties that we liked about two miles apart. All this time we were seeing more of Maria and Louise.

Now what we wanted to do was go trucking. It wasn't long before we had landed ourselves good jobs punching the road trains loaded with logs along State Highway 1. As we carried on thoroughly enjoying this life the weeks turned into months. One Sunday evening the four of us celebrated and raised our glasses to our new way of life in New Zealand.

The following morning Ben and I were a little worse for wear. 'After we're loaded we'll stop in the first lay-by for half an hour's kip,' I said.

Sitting on the bank of lake Taupo I was transfixed by the scenery. It was that beautiful day with the heavenly fresh air smell of the lake mixed with the cut pine on the trucks. While Benny dozed, the trucks made the clicking sound as the engines and exhausts cooled off. I'd asked him whether we had gone too far with Long Shanks. His reply had sent me daydreaming about the period from my army days to now.

When Benny woke he tapped me on the shoulder. 'We'd better get trucking,' he said.

'You were right, Ben. We didn't go too far.'

We made our way back to the rigs and had a quick

look around to make sure everything was in order. They were impressively long – about sixty feet including the drag, The trucks were the latest eight-wheel Fodens, bigger than the English models, each fitted with the latest three-twenty Cummins engine. They had the Jacobs brakes needed in the hilly terrain but hated by townies for the noise they made.

The trucks were fitted with the Fuller Road Range gearboxes and Sheffield-built Kirkstall axles which had two speeds, high and low, that gave the vehicles fifteen gears overall. The exhaust pipe ran up the back of the cab as a stack. Fitted right across the front of the Fodens were large cow catchers.

We thought these rigs – unlike anything we had driven in the UK – were the bees' knees. I was becoming cab happy and I hadn't had that feeling for a long, long time.

We climbed back into our rigs and hit the starter buttons, our engines crackling as they flew into life. As I looked through the huge mirror I could see right down the eight-wheeler and drag. I engaged low third, and as I let the air hand-brake off she gave out a loud hissing sound.

The cab lifted slightly as the engine took the strain and Benny led me back onto the road heading south.

# Other Titles from Old Pond Publishing

## *Trucking and Living in Canada*

DVD aimed at Dutch truckers thinking of emigrating. Interviews, including drivers, police and an instructor, cover working and living conditions.

## *Alaska Highway* TRUCKSTAR

DVD showing owner-operators driving under pressure for timber and oil in extreme conditions from 50 Celsius below to mud and soft going.

## *Heavy Transport* TRUCKSTAR

Seven exceptional loads, mostly of 100–460 tonnes but including a salvaged U-boat flmed in Europe and the United States.

## *Custom Cutters* DYLAN WINTER

Two combine crews filmed through the harvesting season as they head north from Texas towards the Canadian border.

## *Farmer's Boy* MICHAEL HAWKER

A detailed account of farm work in N. Devon in the 1940s and 1950s. Paperback

## *Early to Rise* HUGH BARRETT

A classic of rural literature, this is a truthful account of a young man working as a farm pupil in Suffolk in the 1930s. Paperback

## *A Good Living* HUGH BARRETT

Following on from *Early to Rise*, Hugh takes us back to the assortment of farms with which he was involved from 1937 to 1949. Paperback

## *A Land Girl's War* JOAN SNELLING

Work as a tractor driver on a Norfolk fruit farm and wartime romance vividly recalled. Paperback

## *Land Girls at the Old Rectory* IRENE GRIMWOOD

Light-hearted, boisterous memories of land girls in Suffolk 1942–46. Paperback.

*Free complete catalogue:*

Old Pond Publishing, Dencora Business Centre,
36 White House Road, Ipswich IP1 5LT, United Kingdom
www.oldpond.com    Phone: 01473 238200    Fax: 01473 238201

## About the author

Leslie Purdon was born in Gravesend. His only ambition from infant school onwards was to become a transport driver like other members of his family. During the last months of his school career he played truant in order to haul hardcore in a five-ton Canadian Ford on private land from Gravesend aerodrome to a local building site. After leaving school he became a driver's mate, got his driving licence and was called up for national service.

When Les was demobbed he drove eight-wheel Atkinsons on an A licence, working for Arnold's Transport at Gravesend. He went on to drive for Everard's road tankers, Tankfreight, Pickfords, Thomas Allen and for fourteen years with BP long distance. His final stint was with P & O before he retired in 1998 after 46 years in road transport.

After he retired, Les recalled his early working years in *Truckers North Truckers South*. Much of his spare time now is taken up with his 1946 OLAD Bedford which he and his wife Pauline take to many steam and road truck shows. He lives in Lordswood, Chatham, Kent, where he is known as 'the man with the vintage lorry'.

BC 3|15